AF680078

10 Rhode Island RICAS Grade 8 Math Practice Tests

The Ultimate Test Prep Collection with Answer Explanations

Dr. A. Nazari

Copyright © 2026 Dr. A. Nazari

Published by View Math Education

ViewMath.com

All rights reserved. No part of this publication may be reproduced, distributed, or transmitted in any form or by any means, including photocopying, recording, or other electronic or mechanical methods, without the prior written permission of the author, except in the case of brief quotations embodied in critical reviews and certain other noncommercial uses permitted by copyright law, including Section 107 or 108 of the 1976 United States Copyright Act.

The information in this book is distributed on an "as is" basis, without warranty. While every precaution has been taken in the preparation of this work, neither the author nor the publisher shall have any liability to any person or entity with respect to any loss or damage caused or alleged to be caused directly or indirectly by the information contained in this book.

Copyright © 2026

10 Practice Tests

Grade 8 Mathematics

Welcome!

This book contains ***10 full-length practice tests*** *— the most comprehensive preparation you can get for your Grade 8 math assessment. Each test covers all six topics:*

- *Irrational Numbers*
- *Powers & Scientific Notation*
- *Linear Equations*
- *Functions*
- *Geometry*
- *Data & Relationships*

Ten tests give you the practice needed to walk into the real test feeling fully prepared.

Thorough preparation leads to outstanding results.

“Ten full tests! By the time you finish, there won't be any surprises on test day.”

How to Use This Book

A complete 10-test preparation program

What's Inside

- ***10 Full-Length Practice Tests*** — *each covers all 6 chapters of Grade 8 math: irrational numbers, exponents & scientific notation, linear equations, functions, geometry, and data analysis.*
- ***Detailed Answer Explanations*** — *every question includes a step-by-step solution so you learn from every mistake.*
- ***Formula Reference Sheet*** — *all the key Grade 8 formulas you need, organized and ready for quick review.*
- ***Test Tracker*** — *log your scores across all 10 tests and monitor your progress from start to finish.*

Your 10-Test Training Plan

★ PHASE 1: Foundation (Tests 1–3)

Untimed or soft-timed. Focus on understanding the format, identifying strengths and weaknesses, and building good study habits.

★★ PHASE 2: Building Skills (Tests 4–7)

Timed (70 minutes each). Work on pacing, accuracy, and showing complete solutions. Review weak topics between tests.

★★★ PHASE 3: Test-Day Ready (Tests 8–10)

Full test conditions: strict timing, quiet space, no notes. Compare scores with your early tests to see your growth.

Schedule: *Take one test every 3–4 days, or one per week. Use study days between tests to review.*

Types of Questions

Multiple Choice: *Four options — work the problem first, then match. Eliminate obviously wrong answers to narrow your choices.*

Short Answer & Constructed Response: *Show every step: equations, substitutions, simplifications. Partial credit rewards correct reasoning even if the final answer is off.*

Graphing & Data Analysis: *Plot points, draw lines, interpret graphs. Label axes clearly.*

Tip: Ten tests is a full preparation program. Don't rush. The key is what you do between tests — study, review, and understand your mistakes before moving forward.

Test-Taking Tips

Your complete test-day toolkit

Before the Test

- *Review your notes from the previous test — focus on your weak topics*
- *Set up a quiet, clean workspace with all your materials ready*
- *Start with a positive mindset: you've prepared for this*

During the Test

- *Read each problem fully before calculating anything*
- *Write the formula or set up the equation first, then substitute values*
- *Show all your work — every step, every operation*
- *If stuck for more than 2 minutes, mark it and move on*
- *Use estimation to check if your answers are reasonable*

After the Test

- *Read the full explanation for every question you got wrong*
- *Write down which topics gave you trouble (not just question numbers)*
- *Study those topics before taking the next test*
- *Record your score in the Test Tracker*

Common Mistakes in Grade 8 Math

⚠ ***Exponents:*** $(ab)^n = a^n b^n$, *but* $a^m + a^n \neq a^{m+n}$. *Only multiply/divide to combine.*

⚠ ***Slope formula:*** $m = \frac{y_2 - y_1}{x_2 - x_1}$ *— keep the order consistent.*

⚠ ***Systems of equations:*** *The solution must satisfy both equations.*

⚠ ***Transformations:*** *Rotations and reflections change position; dilations change size.*

⚠ ***Volume:*** *Use* $\pi \approx 3.14$ *or leave as* π *— match what the question asks.*

“The students who improve the most aren't the ones who take the most tests — they're the ones who carefully review every mistake. Make that your priority.”

What You'll Need

Gather these materials before you begin

Materials Checklist

Sharpened Pencils — #2 pencils, at least two

Good Eraser — for clean corrections

Scratch Paper — for working out problems

Ruler / Straightedge — for graphing & geometry

Quiet Space — no distractions

Focused Mind — ready to do your best

Allowed Materials

- ✓ Pencils and eraser
- ✓ Scratch paper (provided on official test day)
- ✓ Ruler or straightedge (if required)
- ✓ Protractor (if required)

Not Allowed

- ✕ Calculator (unless your state test allows it)
- ✕ Cell phone or any electronic device
- ✕ Notes, textbooks, or reference sheets
- ✕ Help from others during the test

A Note for Parents & Guardians

Ten tests is a comprehensive program. Plan ***one test every 3–4 days*** *(or one per week) with study sessions between each test.*

How to help:

- *Tests 1–3 should be untimed — build understanding before adding pressure.*
- *After each test, review the answer explanations together. Ask: "Which topics were hardest? Let's study those before the next one."*
- *Use the Test Tracker to celebrate progress over time.*
- *For topic-specific help, pair this book with our* ***Grade 8 Math Study Guide*** *or* ***Grade 8 Workbook****.*

Grade 8 Formula Reference

Keep this page handy — you may use it during your practice tests!

X^1 Exponent Rules

$a^m \cdot a^n = a^{m+n}$ $(a^m)^n = a^{mn}$ $(ab)^n = a^n \cdot b^n$

$\frac{a^m}{a^n} = a^{m-n}$ $a^0 = 1$ $(a \neq 0)$ $a^{-n} = \frac{1}{a^n}$

Lines & Linear Equations

Slope: $m = \frac{y_2 - y_1}{x_2 - x_1} = \frac{\text{rise}}{\text{run}}$

m = slope b = y-intercept

Slope-intercept: $y = mx + b$

Parallel lines: same slope

Proportional: $y = mx$

Proportional: passes through origin

Scientific Notation

$a \times 10^n$ where $1 \leq |a| < 10$

Multiply: add exponents

Divide: subtract exponents

$\sqrt{x}$ Roots & Number Sense

Perfect squares: 1, 4, 9, 16, 25, 36, 49, 64, 81, 100, 121, 144

Perfect cubes: 1, 8, 27, 64, 125

$\sqrt{2} \approx 1.414$ $\sqrt{3} \approx 1.732$ $\pi \approx 3.14159$

Pythagorean Theorem & Distance

$a^2 + b^2 = c^2$ c = hypotenuse (longest side of a right triangle)

Distance: $d = \sqrt{(x_2 - x_1)^2 + (y_2 - y_1)^2}$

Volume Formulas

Cylinder $V = \pi r^2 h$ **Cone** $V = \frac{1}{3}\pi r^2 h$ **Sphere** $V = \frac{4}{3}\pi r^3$

Angle Relationships

Triangle angle sum: $180°$

Exterior angle = sum of two remote interior angles

Parallel lines + transversal: Alternate interior angles are equal • Co-interior angles sum to $180°$

Functions

Each input → exactly one output

Vertical line test: if any vertical line hits graph more than once ⇒ not a function

Linear: constant rate of change ($y = mx + b$)

Nonlinear: rate of change varies

Transformations

Translation: slide

Reflection: flip

Rotation: turn

Dilation: resize

Congruent = same shape & size

Similar = same shape, proportional size

Tip: Bookmark this page! Review it before each test so these formulas become second nature.

Multiplication Table

You may use this table during your practice tests!

×	1	2	3	4	5	6	7	8	9	10	11	12
1	1	2	3	4	5	6	7	8	9	10	11	12
2	2	4	6	8	10	12	14	16	18	20	22	24
3	3	6	9	12	15	18	21	24	27	30	33	36
4	4	8	12	16	20	24	28	32	36	40	44	48
5	5	10	15	20	25	30	35	40	45	50	55	60
6	6	12	18	24	30	36	42	48	54	60	66	72
7	7	14	21	28	35	42	49	56	63	70	77	84
8	8	16	24	32	40	48	56	64	72	80	88	96
9	9	18	27	36	45	54	63	72	81	90	99	108
10	10	20	30	40	50	60	70	80	90	100	110	120
11	11	22	33	44	55	66	77	88	99	110	121	132
12	12	24	36	48	60	72	84	96	108	120	132	144

How to Use This Table

To find **4** × **7**:

1. Find **4** in the left column (blue).
2. Find **7** in the top row (blue).
3. Follow the row and column until they meet: the answer is **28**!

Tip: You can also use this table for division! If you know $28 \div 4 = ?$, find 28 in the 4's row. The column header gives you the answer: **7***!*

My Test Tracker

Record every test and watch your scores improve

Name: ______________________ **Start Date:** ______________

GETTING STARTED (Tests 1–3)

Test 1 — Untimed

Date: __________ Score: ______/______ %: ______ Topics to review: ____________________

Test 2 — Untimed

Date: __________ Score: ______/______ %: ______ Topics to review: ____________________

Test 3 — Soft Timer

Date: __________ Score: ______/______ %: ______ Topics to review: ____________________

BUILDING SKILLS (Tests 4–7)

Test 4 — Timed (70 min)

Date: __________ Score: ______/______ %: ______ Focus area: ____________________

Test 5 — Timed (70 min)

Date: __________ Score: ______/______ %: ______ Focus area: ____________________

Test 6 — Timed (70 min)

Date: __________ Score: ______/______ %: ______ Focus area: ____________________

Test 7 — Timed (70 min)

Date: __________ Score: ______/______ %: ______ Focus area: ____________________

TEST-DAY READY (Tests 8–10)

Test 8 — Full Test Conditions

Date: ____________ Score: ______/______ %: ______ Growth since Test 1: ____________________

Test 9 — Full Test Conditions

Date: ____________ Score: ______/______ %: ______ Growth since Test 1: ____________________

Test 10 — Full Test Conditions

Date: ____________ Score: ______/______ %: ______ Growth since Test 1: ____________________

Score Progress

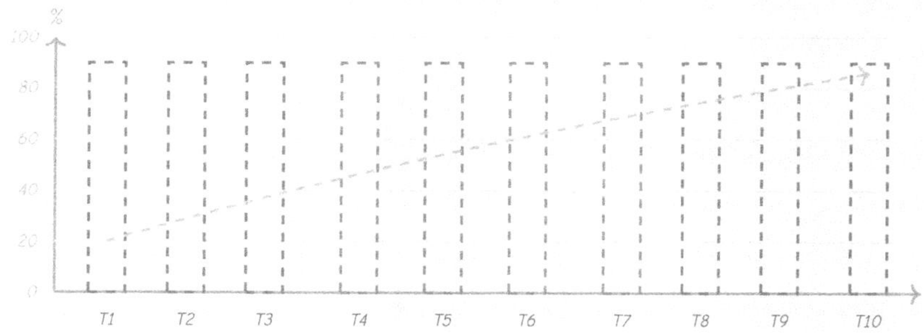

Shade each bar after every test. Watch your improvement!

Final Reflection

The most important thing I learned: ______________________________________

The topic where I improved the most: ______________________________________

My advice for other students: ______________________________________

YOUR ONLINE COMPANION

Continue Learning at ViewMath Academy!

For Parents, Teachers & Students

*Great job on the practice tests! Want to keep improving? ViewMath Academy is your **free online companion** to this book.*

- ***Score Analyzer*** — *Enter your answers and instantly see which topics need more practice*
- ***Interactive Lessons*** — *Review the concepts behind each question with clear explanations*
- ***Adaptive Quizzes*** — *Practice your weak topics with questions that match your level*
- ***Progress Tracking*** — *See your mastery grow across all Grade 8 math topics*
- ***Personalized Dashboard*** — *A learning plan tailored just for you*

Scan to visit ViewMath Academy

ViewMath.com/RI-Grade8

Free to use • No downloads required • Works on any device

Table of Contents

Here's what we'll explore together!

Let's learn and have fun!

PRACTICE TEST

1

Practice Test 1

30 Questions

Before You Start

- **Read each question carefully** before choosing your answer.
- **Show your work** on scratch paper when you need to.
- **Skip hard questions** and come back to them later.
- **Check your answers** when you're done.
- **Take your time** — there's no rush!

You've Got This!

Do your best and show what you know!

1. True or false: The number $\frac{22}{7}$ is equal to π.

Your Answer:

2. The table below shows four repeating decimals and the power of 10 a student used to convert each. Which one uses the WRONG power of 10?

Decimal	Multiply by
$0.\overline{7}$	10
$0.\overline{45}$	100
$0.\overline{123}$	100
$0.\overline{8}$	10

(A) $0.\overline{7}$

(B) $0.\overline{45}$

(C) $0.\overline{123}$

(D) $0.\overline{8}$

3. The table below shows a student's work to approximate $\sqrt{14}$. What should go in the blank?

Guess	Square	Compare to 14
3	9	too small
4	16	too big
3.7	13.69	too small
3.8	14.44	too big
?	?	closest

(A) 3.74, because $3.74^2 = 13.9876$

(B) 3.72, because $3.72^2 = 13.8384$

(C) 3.80, because $3.8^2 = 14.44$

(D) 3.75, because $3.75^2 = 14.0625$

4. *What is the best estimate for* $5 - \sqrt{7}$?

(A) 1.4

(B) 2.4

(C) 3.0

(D) 0.4

5. *The diagram shows a square and a cube. The square has an area of* A *square units and the cube has a volume of* V *cubic units. If* $A = V$*, which pair of side lengths is correct?*

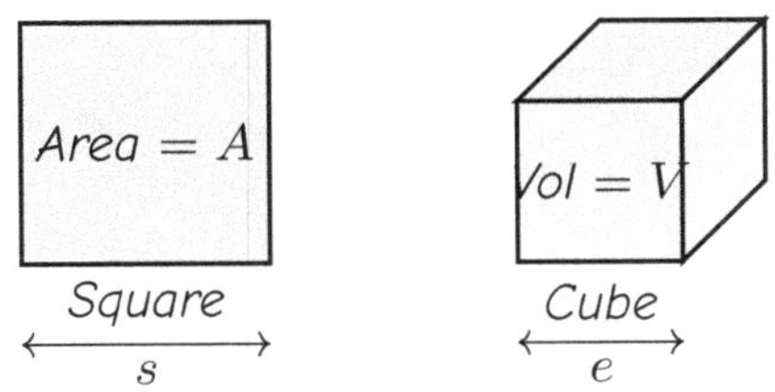

(A) $s = 6,\ e = 6$

(B) $s = 8,\ e = 4$

(C) $s = 9,\ e = 3$

(D) $s = 5,\ e = 5$

6. *Estimate the product* $498{,}000{,}000 \times 6{,}200$ *by first writing each factor in scientific notation and then computing.*

Your Answer:

7. *Study the diagram below. Each box produces its output by multiplying the two inputs. What belongs in the output box marked "?"?*

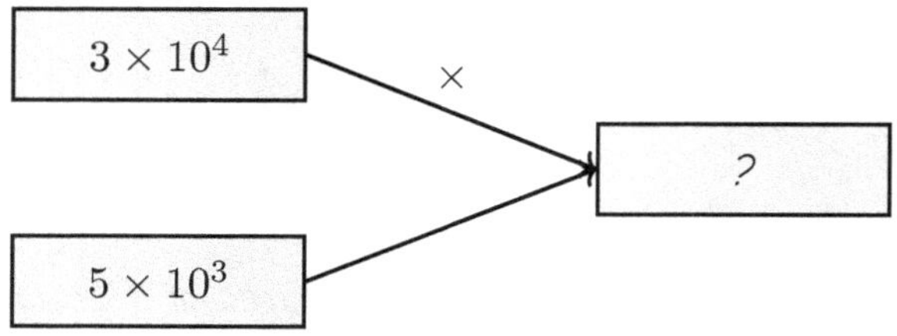

Ⓐ 8×10^7

Ⓑ 1.5×10^7

Ⓒ 1.5×10^8

Ⓓ 15×10^7

8. *A bike rental company charges $\$8$ per hour. Which equation models the total cost y for x hours?*

Ⓐ $y = x + 8$

Ⓑ $y = 8x$

Ⓒ $y = \frac{x}{8}$

Ⓓ $y = 8x + 10$

9. *Find the slope through* $(3, -1)$ *and* $(7, 11)$.

Your Answer:

10. *Solve* $\frac{x+5}{2} = 9$.

Your Answer:

11. *The graph shows the system $y = 2x$ and $y = -x + 9$. At what point do the lines intersect?*

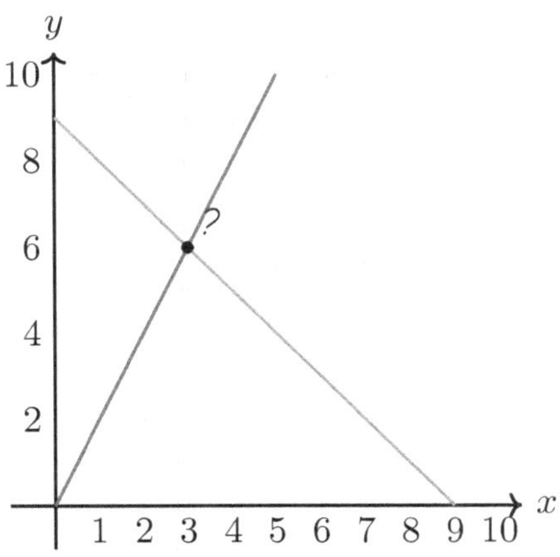

(A) $(2, 4)$

(B) $(3, 6)$

(C) $(4, 8)$

(D) $(5, 10)$

12. *The graph shows a car (starting from home) and a bus (starting 30 miles ahead). The car travels at 50 mph and the bus at 40 mph.*

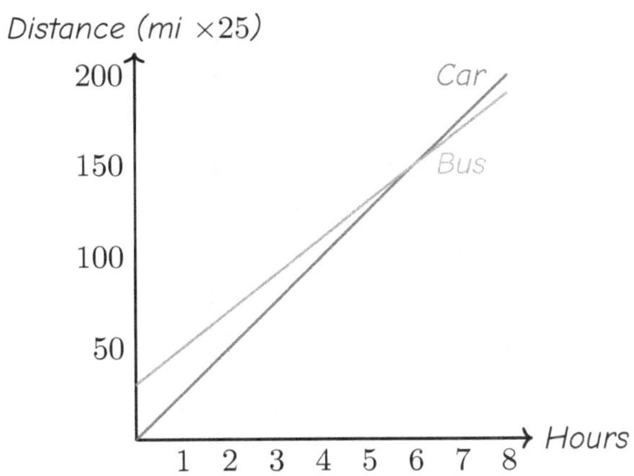

After how many hours does the car catch the bus, and how far from home are they?

Your Answer:

13. How many times can a vertical line cross the graph of a function at most?

Your Answer

14. If $f(x) = 3x + 1$, what is $f(4)$?

(A) 7

(B) 12

(C) 13

(D) 15

15. Company A charges $\$20$ plus $\$8$ per hour. Company B charges $\$50$ plus $\$3$ per hour. After how many hours do they charge the same amount?

Your Answer.

16. Which equation is nonlinear?

(A) $y = -5x + 2$

(B) $y = 10$

(C) $y = 3x$

(D) $y = x^2 - 1$

17. Find the equation of the line through $(1, 3)$ and $(4, 12)$.

Your Answer

18. A graph is flat for 4 seconds on a speed-time graph. What does this mean?

Your Answer.

19. *A parallelogram is reflected over the y-axis. Which of the following changes?*

(A) *The side lengths*

(B) *The angle measures*

(C) *The orientation (left-right order of vertices)*

(D) *The area*

20. *Which statement is true about all congruent figures?*

(A) *They are always in the same orientation.*

(B) *They are always in the same quadrant.*

(C) *One can always be mapped onto the other by rigid transformations.*

(D) *They must be mirror images of each other.*

21. *Which coordinate rule represents a reflection over the x-axis?*

(A) $(x, y) \to (-x, y)$

(B) $(x, y) \to (x, -y)$

(C) $(x, y) \to (-x, -y)$

(D) $(x, y) \to (-y, x)$

22. *Triangle PQR has sides 5, 12, 13. After a dilation by $k = 2$, what is the longest side of tne image?*

(A) 13

(B) 15

(C) 24

(D) 26

23. *Two parallel lines are cut by a transversal. One angle measures $72°$. What is its alternate interior angle?*

(A) $18°$

(B) $72°$

(C) $108°$

(D) $288°$

Get Online
Find more at
ViewMath.com/RI-Grade8
ViewMath.com

24. *An isosceles right triangle has legs of length* 10. *What is the hypotenuse?*

(A) 20

(B) $10\sqrt{2}$

(C) $\sqrt{10}$

(D) 100

25. *What is the distance between* $(-1, 2)$ *and* $(2, 6)$*?*

(A) 25

(B) 7

(C) 5

(D) $\sqrt{13}$

26. *A sphere has the dimensions shown. What is its volume? Use* $\pi \approx 3.14$

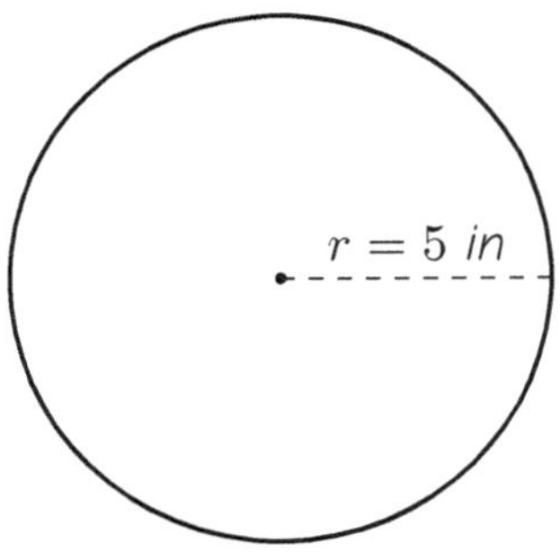

(A) 78.5 in^3

(B) 523.3 in^3

(C) 392.5 in^3

(D) 314.0 in^3

27. *Which of the following is NOT something you should look for when analyzing a scatter plot?*

(A) *Direction (positive, negative, or none)*

(B) *Shape (linear or nonlinear)*

(C) *The exact equation of the data*

(D) *Outliers and clusters*

28. *A trend line passes through* $(2, 10)$ *and* $(8, 4)$. *Find the slope and* y*-intercept.*

Your Answer:

29. *A model is* $y = -0.5x + 25$. *At what value of* x *does the model predict* $y = 0$?

(A) 25

(B) 50

(C) 12.5

(D) -50

30. *Explain the difference between a joint frequency and a marginal frequency.*

Your Answer:

★ End of Practice Test 1 ★

Great job finishing the test!

My Score

I got __________ out of 30 questions right.

*Check your answers in the **Answer Key** at the back of the book.*

Review any questions you missed. That's how we learn!

Check Your Score Online!

*Visit **ViewMath Academy** to enter your answers and see which topics you need to review. You can also explore lessons, take quizzes, track your scores, and save your progress!*

viewmath.com/score/8.1.RI.16

*Or go to **viewmath.com/score** and enter code: **8.1.RI.16***

PRACTICE TEST

2

Practice Test 2

30 Questions

Before You Start

- **Read each question carefully** before choosing your answer.
- **Show your work** on scratch paper when you need to.
- **Skip hard questions** and come back to them later.
- **Check your answers** when you're done.
- **Take your time** — there's no rush!

You've Got This!

Do your best and show what you know!

1. *The number line below shows two points, P and Q.*

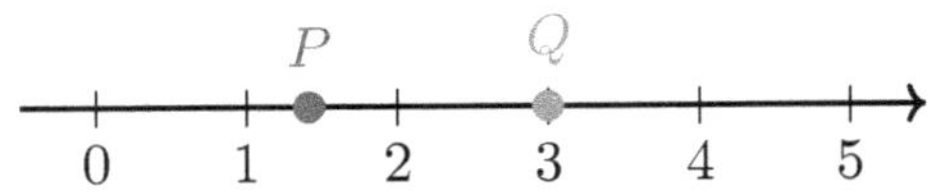

Point P represents $\sqrt{2}$ and point Q represents 3. Which statement is true?

- (A) Both P and Q are rational.
- (B) Both P and Q are irrational.
- (C) P is irrational and Q is rational.
- (D) P is rational and Q is irrational.

2. *The flowchart below shows the conversion steps for a repeating decimal. Fill in the missing values to convert $0.\overline{54}$ to a fraction.*

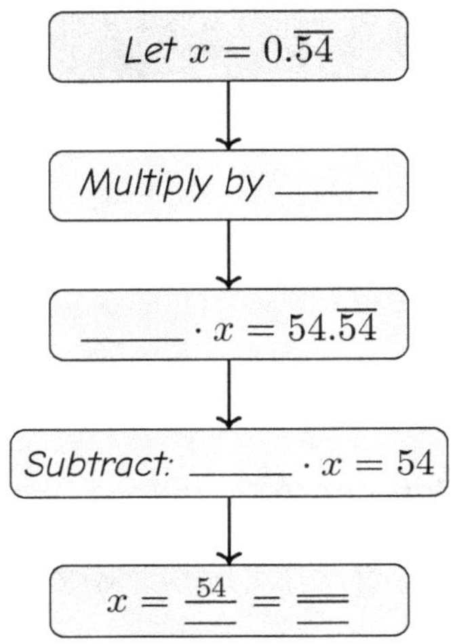

Your Answer:

3. *The squares below have the given areas. Estimate the side length of each square to one decimal place.*

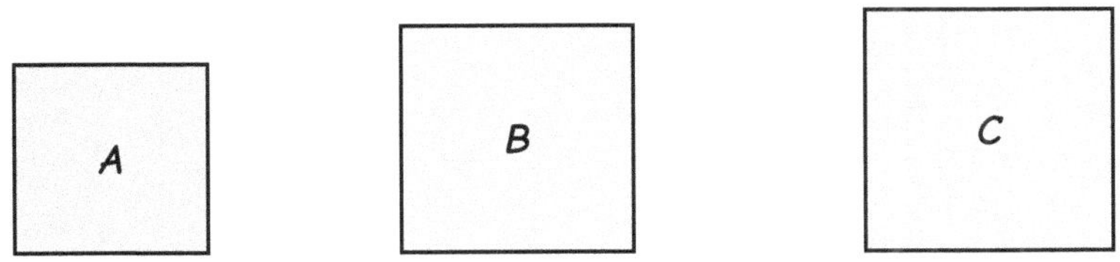

Area $= 18$ *sq units* *Area* $= 40$ *sq units* *Area* $= 72$ *sq units*

Your Answer:

4. *What is the best estimate for* $\sqrt{5}+\sqrt{3}$*?*

(A) $\sqrt{8}\approx 2.8$

(B) 3.0

(C) 4.0

(D) 8.0

5. *Solve* $x^2 = \frac{9}{16}$.

(A) $x = \frac{3}{4}$ *only*

(B) $x = \pm\frac{3}{4}$

(C) $x = \frac{9}{8}$

(D) $x = \pm\frac{9}{4}$

6. *Which of the following is true about* 4.2×10^{-3} *and* 4.2×10^{3}*?*

(A) *They are equal*

(B) 4.2×10^{3} *is* 10^{6} *times as large as* 4.2×10^{-3}

(C) 4.2×10^{-3} *is* 10^{6} *times as large as* 4.2×10^{3}

(D) *Their product is* 1

7. *What is* $\frac{8\times 10^{7}}{2\times 10^{3}}$*?*

(A) 4×10^{4}

(B) 6×10^{4}

(C) 4×10^{10}

(D) 16×10^{10}

Get Online
Find more at
ViewMath.com/RI-Grade8
ViewMath.com

8. A faucet leaks at a constant rate. It leaks 45 mL in 15 minutes. How many mL does it leak in 1 hour?

Your Answer

9. A plant grows from 4 cm to 16 cm in 6 weeks. What is the rate of change?

(A) 1 cm/week
(B) 2 cm/week
(C) 3 cm/week
(D) 4 cm/week

10. How many solutions does $2(x+3) = 2x+6$ have?

Your Answer

11. Solve by substitution: $x = y + 2$ and $3x + y = 14$.

(A) $(3, 1)$
(B) $(4, 2)$
(C) $(5, 3)$
(D) $(6, 4)$

12. A canoe goes 18 km downstream in 2 hours and 18 km upstream in 3 hours. What is the canoe's speed in still water?

Your Answer

13. Look at the graph of the relation below. Explain whether it is a function and why.

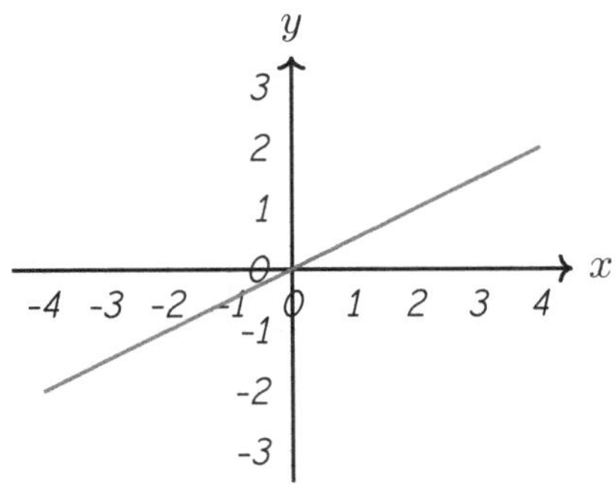

Your Answer:

14. If $h(x) = -3x + 15$, for what value of x is $h(x) = 0$?

Your Answer:

15. Function A starts at \$50 and increases by \$12 per week. Function B: $y = 15x + 30$. Which function starts with more money?

- (A) Function A
- (B) Function B
- (C) They start with the same amount.
- (D) Cannot be determined

16. A table shows $(0, 3)$, $(1, 7)$, $(2, 11)$, $(3, 15)$. Is this linear or nonlinear?

- (A) Nonlinear, because the outputs are odd numbers only.
- (B) Linear, because the change in y is always 4.
- (C) Nonlinear, because the outputs get larger.
- (D) Linear, because the first output is positive.

17. The function $y = 50x + 200$ models a bank account balance after x weeks. What is the weekly deposit?

(A) \$50
(B) \$100
(C) \$200
(D) \$250

18. A graph shows temperature over a day. From 6 AM to noon, the graph goes up steeply. From noon to 3 PM it goes up slowly. What can you conclude?

(A) The temperature decreased between noon and 3 PM.
(B) The temperature increased faster in the morning than in the afternoon.
(C) The temperature was constant from noon to 3 PM.
(D) The temperature increased at the same rate all day.

19. A triangle is translated 6 units left and 3 units up. A vertex was at $(2, -1)$. What are its new coordinates?

Your Answer:

20. Which pair of figures is NOT necessarily congruent?

(A) A triangle and its reflection
(B) A rectangle and its translation
(C) Two rectangles with the same perimeter
(D) A square and its $90°$ rotation

21. A point $(-1, 6)$ is rotated $90°$ clockwise around the origin. The $90°$ clockwise rule is $(x, y) \to (y, -x)$. What is the image?

(A) $(6, 1)$
(B) $(-6, -1)$
(C) $(-6, 1)$
(D) $(1, -6)$

22. True or false: All rectangles are similar to each other.

Your Answer

23. Two parallel lines are cut by a transversal. If one co-interior (same-side interior) angle is $135°$, what is the other?

(A) $135°$
(B) $45°$
(C) $55°$
(D) $225°$

24. A right triangle has legs a and $b = 4a$. If the hypotenuse is $\sqrt{17} \cdot a$, verify using the Pythagorean Theorem.

Your Answer

25. Points P and Q are shown on the grid. What is the distance between them?

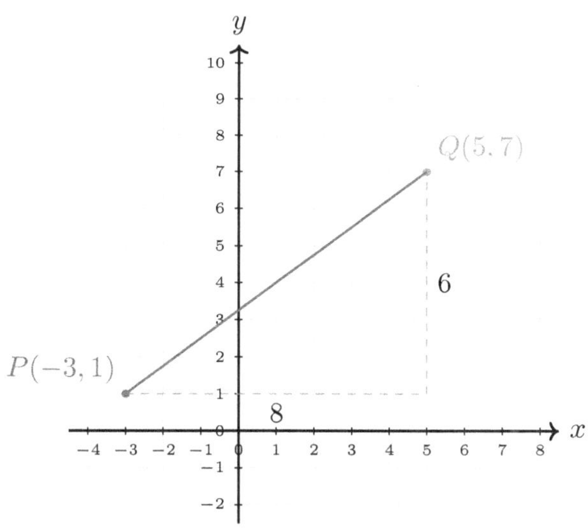

(A) 14
(B) $\sqrt{14}$
(C) 10
(D) $\sqrt{28}$

26. *A waffle cone is shaped like a cone with radius 2.5 cm and height 12 cm. A spherical scoop of ice cream with radius 2.5 cm sits on top (as a half-sphere). Find the total volume. Use $\pi \approx 3.14$.*

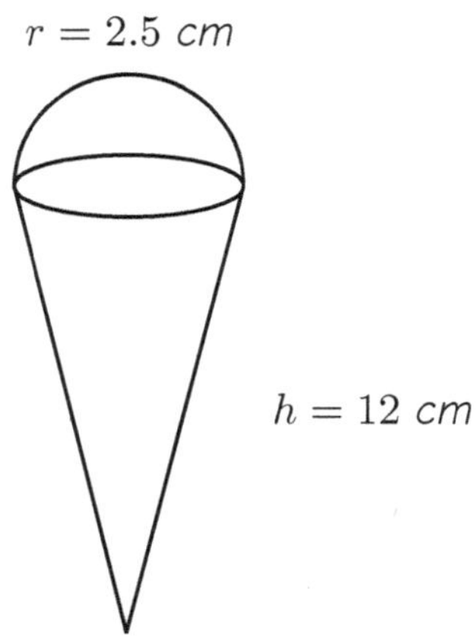

Your Answer:

27. *A scatter plot shows that as the number of hours practiced increases, the number of free throws made also increases. What type of association is this?*

(A) Negative association
(B) No association
(C) Positive association
(D) Nonlinear association

28. *Data: $(1,3), (2,5), (3,4), (4,6), (5,8)$. A trend line through $(1,3)$ and $(5,8)$ has slope:*

(A) $\frac{5}{4}$
(B) $\frac{4}{5}$
(C) 1
(D) 2

29. *A scatter plot shows data and a trend line. The actual point at $x = 3$ is at $y = 8$, but the trend line passes through $(3, 6)$. What is the residual?*

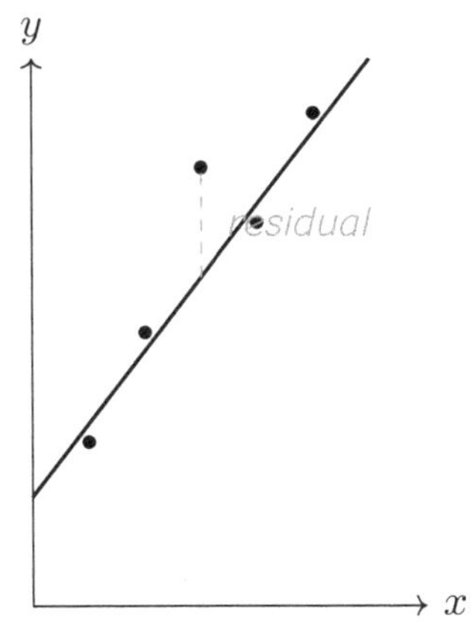

(A) -2

(B) 2

(C) 6

(D) 14

30. *A segmented (stacked) bar chart shows the proportion of students choosing band or chorus by gender. The boys' bar is 60% band and 40% chorus. The girls' bar is 45% band and 55% chorus. What can you conclude?*

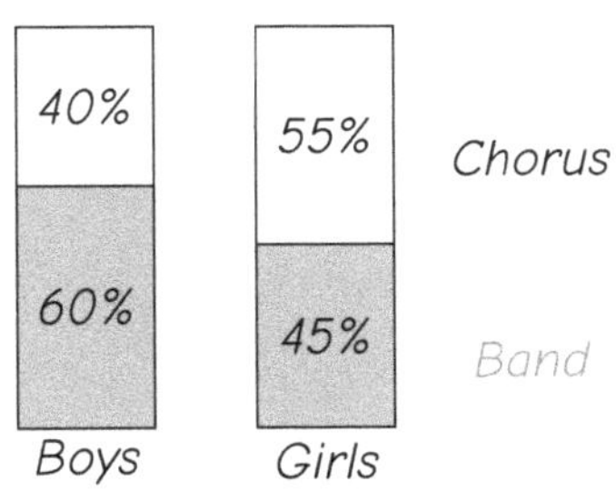

(A) *There is no association between gender and music choice.*

(B) *Boys are more likely to choose band; girls more likely to choose chorus.*

(C) *Girls and boys prefer band equally.*

(D) *Chorus is more popular overall.*

★ End of Practice Test 2 ★

Great job finishing the test!

My Score

I got __________ out of 30 questions right.

*Check your answers in the **Answer Key** at the back of the book.*

Review any questions you missed. That's how we learn!

Check Your Score Online!

*Visit **ViewMath Academy** to enter your answers and see which topics you need to review. You can also explore lessons, take quizzes, track your scores, and save your progress!*

viewmath.com/score/8.1.RI.17

*Or go to **viewmath.com/score** and enter code: **8.1.RI.17***

PRACTICE TEST

3

Practice Test 3

30 Questions

Before You Start

- **Read each question carefully** before choosing your answer.
- **Show your work** on scratch paper when you need to.
- **Skip hard questions** and come back to them later.
- **Check your answers** when you're done.
- **Take your time** — there's no rush!

You've Got This!

Do your best and show what you know!

1. Which number below can be written as a fraction $\frac{a}{b}$ where a and b are integers and $b \neq 0$?

(A) $\sqrt{3}$

(B) π

(C) $0.\overline{81}$

(D) $\sqrt{11}$

2. What is $0.\overline{27}$ as a fraction in simplest form?

(A) $\frac{27}{100}$

(B) $\frac{27}{99}$

(C) $\frac{3}{11}$

(D) $\frac{9}{33}$

3. A student estimated $\sqrt{60} \approx 8$. Is this estimate reasonable?

(A) Yes, because $8^2 = 64$ is close to 60.

(B) No, because $8^2 = 64$, which is too far from 60. A better estimate is about 7.7.

(C) No, $\sqrt{60} = 30$.

(D) Yes, because $60 \div 8 \approx 8$.

4. Which is larger: $4\sqrt{2}$ or $\sqrt{30}$?

(A) $4\sqrt{2}$, because $4 > \sqrt{30}$

(B) $\sqrt{30}$, because $30 > 2$

(C) $4\sqrt{2}$, because $4\sqrt{2} \approx 5.66 > 5.48 \approx \sqrt{30}$

(D) They are equal.

5. A square has an area of 196 square centimeters. What is the length of one side?

(A) 13 cm

(B) 49 cm

(C) 14 cm

(D) 98 cm

6. Which of the following is 0.00072 written in scientific notation?

(A) 7.2×10^{-4}

(B) 7.2×10^{4}

(C) 72×10^{-5}

(D) 0.72×10^{-3}

7. A virus has a mass of 9.5×10^{-18} grams. If a sample contains 2×10^{6} viruses, what is the total mass?

(A) 1.9×10^{-11} g

(B) 1.9×10^{-12} g

(C) 11.5×10^{-12} g

(D) 1.9×10^{-24} g

8. Store A sells apples for \$2 per pound. Store B's prices are shown in the table.

Pounds	3	6	9
Cost (\$)	7.50	15.00	22.50

Which store has the lower unit price?

(A) Store A

(B) Store B

(C) They charge the same price.

(D) Not enough information.

9. A car's odometer reads 120 miles at 2:00 PM and 270 miles at 5:00 PM. What is the car's speed (rate of change)?

(A) 45 mph

(B) 50 mph

(C) 60 mph

(D) 90 mph

10. Solve $5(x+2) = 5x + 10$.

(A) $x = 0$

(B) $x = 2$

(C) No solution

(D) Infinitely many solutions

11. Solve: $y = \frac{1}{2}x + 3$ and $y = 2x - 3$.

(A) $(2, 4)$

(B) $(4, 5)$

(C) $(6, 6)$

(D) $(8, 7)$

Get Online
Find more at
ViewMath.com/RI-Grade8

ViewMath.com

12. One number is 4 more than another. Their sum is 36. Find both numbers.

Your Answer:

13. The table below shows a relation. Is it a function?

x	1	2	3	4
y	5	10	15	20

(A) No, because the outputs are too large.

(B) No, because the inputs are consecutive.

(C) Yes, because each input has exactly one output.

(D) Yes, because the outputs are all different.

14. If $g(x) = x^2 + 3$, find $g(-2)$.

Your Answer:

15. Function A: $y = 2x + 100$. Function B: $y = 8x + 10$. At what value of x do the functions have the same output?

(A) $x = 10$

(B) $x = 15$

(C) $x = 18$

(D) $x = 20$

16. A table shows $(1, 3)$, $(2, 6)$, $(3, 11)$, $(4, 18)$. Is the function linear or nonlinear?

(A) Linear, because y increases as x increases.

(B) Linear, because the first difference is 3.

(C) Nonlinear, because the differences in y are $3, 5, 7$ — not constant.

(D) Nonlinear, because all outputs are positive.

17. The table below represents a linear function. What is the equation?

x	0	1	2	3	4
y	7	10	13	16	19

(A) $y = 7x + 3$

(B) $y = 3x + 10$

(C) $y = 3x + 7$

(D) $y = 10x + 7$

18. Which scenario matches a graph that is decreasing and linear?

(A) A bank account earning interest

(B) A car driving at constant speed losing fuel at a steady rate

(C) A ball bouncing up and down

(D) A tree growing over many years

19. Point $R(-1, 4)$ is rotated $90°$ counterclockwise around the origin. What are the coordinates of R'?

(A) $(4, 1)$

(B) $(-4, -1)$

(C) $(1, -4)$

(D) $(-4, 1)$

20. Triangle RST has vertices $R(0,0)$, $S(4,0)$, and $T(0,3)$. Triangle $R'S'T'$ has vertices $R'(0,0)$, $S'(0,4)$, and $T'(-3,0)$. What transformation maps RST to $R'S'T'$?

(A) Translation

(B) Reflection over the x-axis

(C) $90°$ counterclockwise rotation

(D) $90°$ clockwise rotation

21. A point is dilated by factor 3 from the origin. If the image is $(9, -15)$, what was the original point?

(A) $(27, -45)$

(B) $(3, -5)$

(C) $(6, -12)$

(D) $(12, -18)$

22. *Triangle A has sides $3, 4, 5$ as shown. Triangle B is similar to A with a scale factor of 2. What is the hypotenuse of Triangle B?*

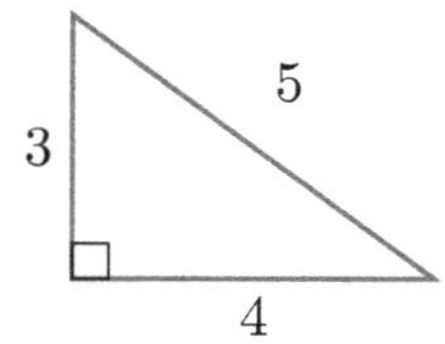

Triangle A

Your Answer:

23. *Parallel lines ℓ and m are cut by transversal t. If a pair of alternate interior angles are $(4x)^\circ$ and 80°, what is x?*

Ⓐ 320 Ⓑ 20

Ⓒ 25 Ⓓ 10

24. *A rectangular TV screen is 36 inches wide and 27 inches tall. What is the diagonal length?*

Ⓐ 63 *in* Ⓑ 45 *in*

Ⓒ $\sqrt{63}$ *in* Ⓓ 54 *in*

25. *Two corners of a square are at $(0,0)$ and $(0,6)$. What is the length of the diagonal?*

Ⓐ 6 Ⓑ 12

Ⓒ $6\sqrt{2}$ Ⓓ 36

26. *A cone has volume 100 cm^3. A cylinder with the same base and height has volume ___ cm^3.*

Your Answer:

27. A scatter plot has a strong upward trend. Which r-value (correlation) is most likely?

(A) $r = -0.9$
(B) $r = 0.1$
(C) $r = 0.9$
(D) $r = 0$

28. A line passes through $(1, 4)$ and $(5, 12)$. What is the slope?

(A) 4
(B) 2
(C) 8
(D) $\frac{1}{2}$

29. A model predicts that for every additional year of age, a plant grows 1.5 cm. This 1.5 is the:

(A) y-intercept
(B) slope
(C) x-intercept
(D) correlation

30.

	Pizza	Tacos	Total
Boys	24	16	40
Girls	18	22	40
Total	42	38	80

What is the relative frequency of boys who prefer pizza out of all 80 students?

(A) 0.30
(B) 0.60
(C) 0.20
(D) 0.525

End of Practice Test 3

Great job finishing the test!

My Score

I got __________ out of 30 questions right.

*Check your answers in the **Answer Key** at the back of the book.*

Review any questions you missed. That's how we learn!

Check Your Score Online!

*Visit **ViewMath Academy** to enter your answers and see which topics you need to review. You can also explore lessons, take quizzes, track your scores, and save your progress!*

viewmath.com/score/8.1.RI.18

*Or go to **viewmath.com/score** and enter code: **8.1.RI.18***

PRACTICE TEST

4

Practice Test 4

30 Questions

Before You Start

- **Read each question carefully** before choosing your answer.
- **Show your work** on scratch paper when you need to.
- **Skip hard questions** and come back to them later.
- **Check your answers** when you're done.
- **Take your time** — there's no rush!

You've Got This!

Do your best and show what you know!

1. *Which of the following numbers is irrational?*

   (A) $\frac{5}{8}$
   (B) $0.\overline{6}$
   (C) $\sqrt{9}$
   (D) $\sqrt{7}$

2. *Write* $0.5\overline{8}$ *as a fraction in simplest form.*

   Your Answer:

3. *Which inequality is true?*

   (A) $\sqrt{15} < 3$
   (B) $\sqrt{15} > 4$
   (C) $\sqrt{15} > 3$
   (D) $\sqrt{15} = 4$

4. *Compute* $\sqrt{2} \times \sqrt{18}$ *exactly, without a calculator.*

   Your Answer:

5. *What is* $\sqrt{49}$*?*

   (A) 6
   (B) 7
   (C) 8
   (D) 24.5

6. *What is* 6.1×10^3 *written in standard form?*

   (A) 61,000
   (B) 610
   (C) 6,100
   (D) 0.0061

7. *A bacterium has a length of* 2×10^{-6} *m. If* 1 *micrometer (*μm*)* $= 10^{-6}$ *m, what is the bacterium's length in micrometers?*

(A) 2×10^{-12} μm

(B) 0.002 μm

(C) 2 μm

(D) $2{,}000$ μm

8. *Taxi A charges* $\$3$ *per mile. Taxi B charges* $\$36$ *for a* 9*-mile trip. Which taxi has the higher unit rate?*

Your Answer:

9. *What is the slope through* $(-4, -1)$ *and* $(2, 5)$*?*

(A) -1

(B) $\frac{2}{3}$

(C) 1

(D) $\frac{3}{2}$

10. *Solve* $5x - 3 = 2x + 9$*.*

(A) $x = 2$

(B) $x = 3$

(C) $x = 4$

(D) $x = 6$

11. *How many solutions does this system have?* $y = 2x + 5$ *and* $4x - 2y = -10$*.*

(A) *No solution*

(B) *One solution*

(C) *Two solutions*

(D) *Infinitely many solutions*

12. *You have* $\$5$ *bills and* $\$10$ *bills totaling* $\$85$*. You have* 12 *bills. How many* $\$10$ *bills do you have?*

Your Answer:

13. Which graph below represents a function?

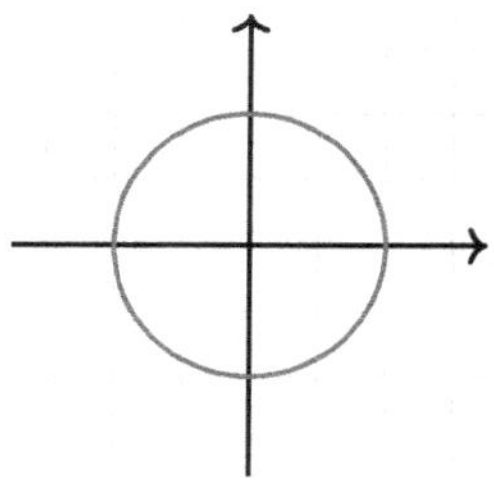

Graph A

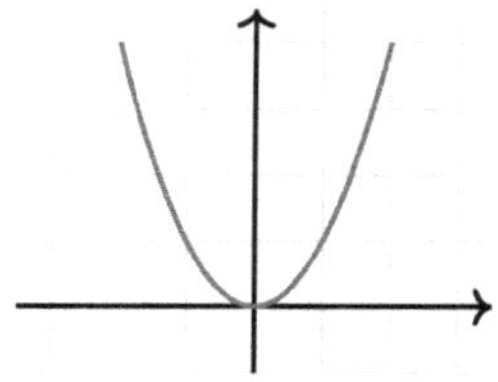

Graph B

Ⓐ Graph A only

Ⓑ Graph B only

Ⓒ Both Graph A and Graph B

Ⓓ Neither Graph A nor Graph B

14. A function table shows $f(1) = 4$, $f(2) = 7$, $f(3) = 10$. What is $f(2) - f(1)$?

Your Answer

15. Function A starts at \$50 and increases by \$12 per week. Function B: $y = 15x + 30$. Which earns more per week?

Ⓐ Function A

Ⓑ Function B

Ⓒ They earn the same per week.

Ⓓ Cannot be determined.

16. *Which graph shows a function with a constant rate of change?*

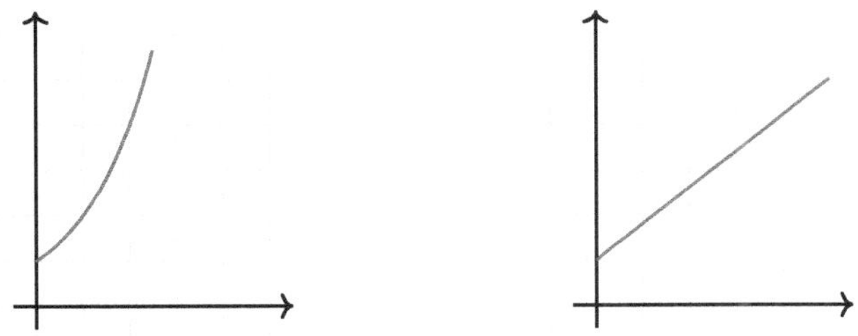

Graph P *Graph Q*

(A) *Graph P*

(B) *Graph Q*

(C) *Both*

(D) *Neither*

17. *A phone plan costs* $35 *per month with no activation fee. Which equation models the total cost?*

(A) $y = 35$

(B) $y = 35x$

(C) $y = 35x + 35$

(D) $y = x + 35$

18. *Which graph could represent a person standing still for* 5 *seconds, then walking at a steady pace?*

(A) *A line sloping upward for the entire time*

(B) *A horizontal line, then a line sloping upward*

(C) *A line sloping downward, then a horizontal line*

(D) *A curve going upward the entire time*

19. *Which transformation turns a figure around a fixed point?*

(A) *Translation*

(B) *Reflection*

(C) *Rotation*

(D) *Dilation*

Get Online

Find more at
ViewMath.com/RI-Grade8

ViewMath.com

20. *Which transformation would NOT preserve congruence?*

(A) *Reflection over the x-axis*

(B) *Translation 2 units up*

(C) *Dilation by a factor of 1.5*

(D) *Rotation of 270°*

21. *Point $(-3, 5)$ is rotated $90°$ counterclockwise around the origin. What is its image?*

(A) $(5, 3)$

(B) $(-5, -3)$

(C) $(3, -5)$

(D) $(-5, 3)$

22. *A dilation with $k = 1$ produces:*

(A) *A figure twice as large*

(B) *A congruent figure*

(C) *A figure half as large*

(D) *No figure at all*

23. *Two parallel lines are cut by a transversal as shown. What is the value of x?*

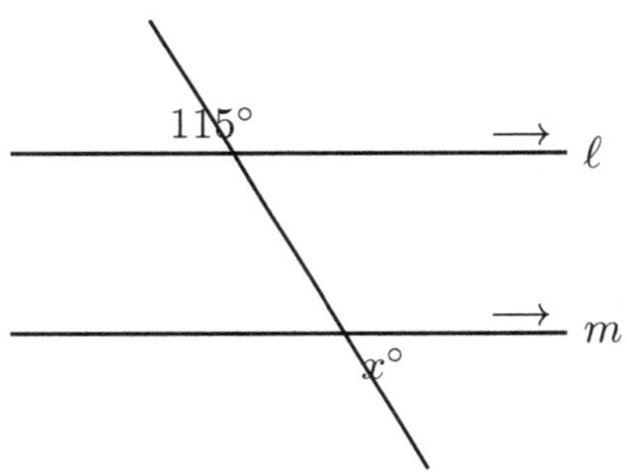

(A) 115°

(B) 65°

(C) 25°

(D) 180°

24. A right triangle has legs 10 and 24. What is the hypotenuse?

Your Answer:

25. What is the distance between $(1, 3)$ and $(4, 7)$?

(A) 7
(B) 25
(C) $\sqrt{7}$
(D) 5

26. A cylinder has radius r and height h. If the radius is doubled, how does the volume change?

(A) It doubles.
(B) It triples.
(C) It quadruples (becomes 4 times as large).
(D) It becomes 8 times as large.

27. True or false: A scatter plot can show both clustering and outliers at the same time.

(A) True — they describe different features of the data.
(B) False — data has either clusters or outliers, not both.
(C) True — but only if there is no association.
(D) False — outliers are always in clusters.

28. A trend line passes through $(2, 5)$ and $(6, 13)$. What is the equation of the line?

(A) $y = 2x + 1$
(B) $y = 2x + 5$
(C) $y = 4x - 3$
(D) $y = x + 3$

Get Online
Find more at
ViewMath.com/RI-Grade8
ViewMath.com

29. A model for plant height is $y = 0.4x + 2$ where x is days and y is height in cm. Using the graph, find the predicted height at day 10, and explain what the slope and y-intercept mean.

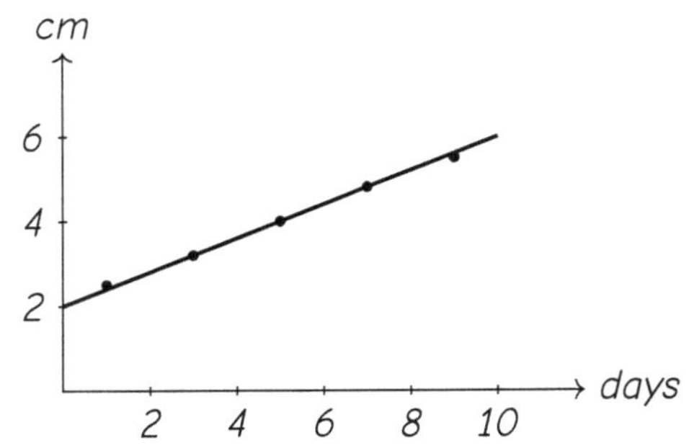

Your Answer:

30. A relative frequency table shows the value 0.15 in one cell. If the total is 200, what is the frequency?

(A) 15

(B) 30

(C) 10

(D) 20

★ End of Practice Test 4 ★

Great job finishing the test!

My Score

I got __________ out of 30 questions right.

*Check your answers in the **Answer Key** at the back of the book.*

Review any questions you missed. That's how we learn!

Check Your Score Online!

*Visit **ViewMath Academy** to enter your answers and see which topics you need to review. You can also explore lessons, take quizzes, track your scores, and save your progress!*

viewmath.com/score/8.1.RI.19

*Or go to **viewmath.com/score** and enter code: 8.1.RI.19*

PRACTICE TEST

5

Practice Test 5

30 Questions

Before You Start

- **Read each question carefully** before choosing your answer.
- **Show your work** on scratch paper when you need to.
- **Skip hard questions** and come back to them later.
- **Check your answers** when you're done.
- **Take your time** — there's no rush!

You've Got This!

Do your best and show what you know!

1. *A calculator shows* $\sqrt{144} = 12$. *Is* $\sqrt{144}$ *rational or irrational? Explain.*

Your Answer

2. *To convert* $0.\overline{63}$ *to a fraction, which equation should you set up after letting* $x = 0.\overline{63}$?

(A) $10x - x = 63$

(B) $100x - x = 63$

(C) $100x - x = 6.3$

(D) $1000x - x = 63$

3. *Which inequality correctly describes the location of* $\sqrt{22}$?

(A) $4 < \sqrt{22} < 5$

(B) $5 < \sqrt{22} < 6$

(C) $10 < \sqrt{22} < 12$

(D) $\sqrt{22} = 11$

4. *The diagram below shows a rectangle with irrational side lengths. Which is the best estimate of the area?*

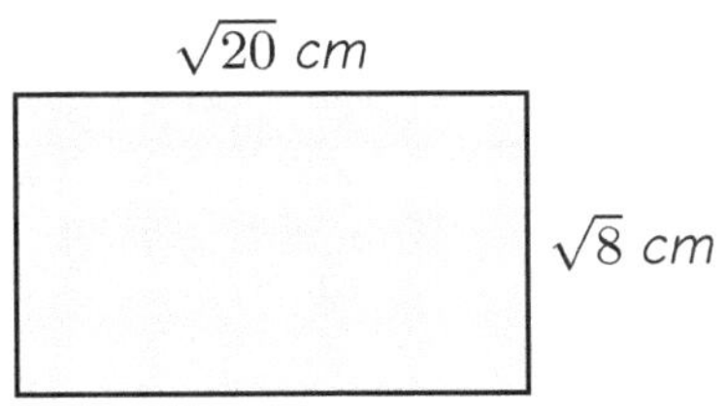

(A) $\sqrt{28} \approx 5.3$ cm^2

(B) $\sqrt{160} \approx 12.6$ cm^2

(C) 28 cm^2

(D) 160 cm^2

5. *Which of the following is a perfect square?*

(A) 50

(B) 72

(C) 81

(D) 90

Get Online
Find more at
ViewMath.com/RI-Grade8
ViewMath.com

6. *A hydrogen atom has a radius of about* 2.5×10^{-11} *meters. How many places do you move the decimal to write this in standard form?*

Your Answer

7. *What is* $5.2 \times 10^6 + 3.8 \times 10^6$*?*

(A) 9×10^6

(B) 9×10^{12}

(C) 19.76×10^6

(D) 9×10^{36}

8. *Which ordered pair could NOT lie on the graph of a proportional relationship?*

(A) $(0, 0)$

(B) $(1, 5)$

(C) $(3, 15)$

(D) $(2, 13)$

9. *Find the slope through* $(\frac{1}{2}, 3)$ *and* $(\frac{3}{2}, 7)$.

(A) 2

(B) 4

(C) $\frac{1}{4}$

(D) 8

10. *Solve* $3(2x - 5) = 4x + 1$.

Your Answer

11. *Solve:* $y = 2x + 1$ *and* $y = -x + 10$.

Your Answer

12. Pens cost \$2 and notebooks cost \$5. You buy 8 items for \$25. How many notebooks did you buy?

(A) 2
(B) 3
(C) 4
(D) 5

13. Give an example of a set of three ordered pairs that is a function.

Your Answer

14. A function is defined by the table below.

x	0	1	2	3
$f(x)$	-1	3	7	11

What is $f(0) + f(2)$?

(A) 2
(B) 6
(C) 8
(D) 10

15. Function A: $y = 6x$. Function B passes through $(1, 5)$ and $(3, 15)$. Which has a greater rate of change?

(A) Function A
(B) Function B
(C) They have the same rate of change.
(D) Cannot be determined.

16. A function table shows $(0, 2)$, $(1, 6)$, $(2, 10)$, $(3, 14)$. What is the constant rate of change?

Your Answer

Get Online
Find more at
ViewMath.com/RI-Grade8
ViewMath.com

17. *A line passes through* $(-1, 2)$ *and* $(3, 10)$. *What is the slope?*

(A) 2

(B) 3

(C) 4

(D) 8

18. *Which word describes a graph that goes downward from left to right?*

(A) *Increasing*

(B) *Constant*

(C) *Decreasing*

(D) *Linear*

19. *A point at* $(-4, 3)$ *is rotated* $180°$ *around the origin. What are the new coordinates?*

(A) $(4, 3)$

(B) $(-4, -3)$

(C) $(4, -3)$

(D) $(3, -4)$

20. *Rectangle* $ABCD$ *has* $AB = 6$ *and* $BC = 10$. *Rectangle* $WXYZ$ *has* $WX = 10$ *and* $XY = 6$. *Are they congruent?*

(A) *No, because the side lengths are in different order.*

(B) *No, because* $6 \neq 10$.

(C) *Yes, because the sets of side lengths are the same.*

(D) *Cannot be determined.*

21. *Dilate the point* $(-4, 10)$ *by a factor of* $\frac{1}{2}$ *from the origin. What is the image?*

Your Answer:

22. *A rectangle is 3 cm by 5 cm. Another rectangle is 6 cm by 9 cm. Are they similar?*

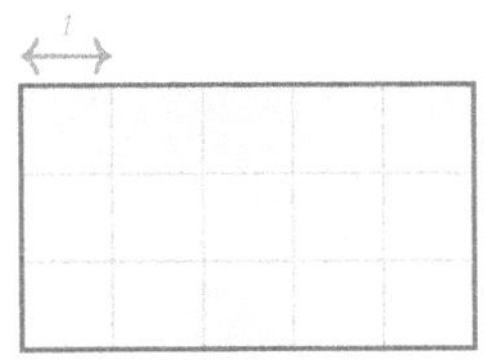

- (A) *Yes, with scale factor* 2
- (B) *Yes, with scale factor* 3
- (C) *No, the sides are not proportional.*
- (D) *Yes, with scale factor* $\frac{3}{5}$

23. *Two angles of a triangle are* $37°$ *and* $53°$. *What is the exterior angle adjacent to the third angle?*

- (A) $90°$
- (B) $127°$
- (C) $53°$
- (D) $143°$

24. *In a right triangle, the legs are* 6 *and* 8. *What is the hypotenuse?*

- (A) 14
- (B) 10
- (C) 48
- (D) $\sqrt{48}$

25. *What is the distance between* $(-3,-4)$ *and* $(0,0)$*?*

- (A) 7
- (B) $\sqrt{7}$
- (C) 5
- (D) 25

26. *A basketball has a diameter of* 24 *cm. What is its volume? Use* $\pi \approx 3.14$

- (A) 7,234.6 cm^3
- (B) 2,143.6 cm^3
- (C) 904.3 cm^3
- (D) 18,086.4 cm^3

27. *What three things should you describe when analyzing a scatter plot?*

Your Answer:

28. *If two students draw different lines of best fit for the same data, which is better?*

Ⓐ The one with the steeper slope

Ⓑ The one where all points are above the line

Ⓒ The one that better balances points above and below and stays close to the data

Ⓓ Both lines are always equally good

29. *Use the trend line $y = 1.5x + 2$ shown below. Predict the value of y at $x = 6$ and calculate the residual if the actual value is $y = 10$.*

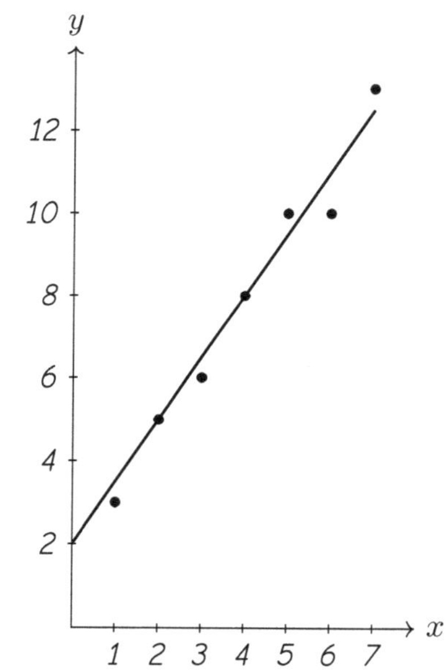

Your Answer:

30. *Using the table above, what proportion of pizza lovers are boys?*

(A) $\frac{18}{42}$

(B) $\frac{24}{42}$

(C) $\frac{24}{80}$

(D) $\frac{24}{40}$

End of Practice Test 5

Great job finishing the test!

My Score

I got __________ out of 30 questions right.

Check your answers in the ***Answer Key*** *at the back of the book.*

Review any questions you missed. That's how we learn!

Check Your Score Online!

Visit ***ViewMath Academy*** *to enter your answers and see which topics you need to review. You can also explore lessons, take quizzes, track your scores, and save your progress!*

viewmath.com/score/8.1.RI.20

Or go to **viewmath.com/score** *and enter code:* **8.1.RI.20**

PRACTICE TEST

6

Practice Test 6

30 Questions

Before You Start

- **Read each question carefully** before choosing your answer.
- **Show your work** on scratch paper when you need to.
- **Skip hard questions** and come back to them later.
- **Check your answers** when you're done.
- **Take your time** — there's no rush!

You've Got This!

Do your best and show what you know!

1. *Is $\frac{5}{6}$ rational or irrational? Explain.*

Your Answer

2. *When converting $0.\overline{123}$ to a fraction, by what power of 10 should you multiply both sides?*

(A) 10
(B) 100
(C) 1000
(D) 10000

3. *Between which two consecutive integers does $\sqrt{110}$ lie?*

(A) 9 *and* 10
(B) 10 *and* 11
(C) 11 *and* 12
(D) 54 *and* 56

4. *A student says $\sqrt{4}+\sqrt{9}=\sqrt{13}$. Is this correct?*

(A) *Yes, because you add the numbers under the square roots.*
(B) *No,* $\sqrt{4}+\sqrt{9}=2+3=5$, *but* $\sqrt{13}\approx 3.6$.
(C) *Yes, because* $4+9=13$.
(D) *No,* $\sqrt{4}+\sqrt{9}=\sqrt{36}=6$.

5. *A cube has a volume of 343 cubic inches. What is the length of one edge?*

(A) 7 *in.*
(B) 49 *in.*
(C) $114.\overline{3}$ *in.*
(D) 17 *in.*

6. *What is 9.03×10^{-5} written in standard form?*

(A) $903{,}000$
(B) 0.0000903
(C) 0.000903
(D) 0.00903

Get Online
Find more at
ViewMath.com/RI-Grade8

ViewMath.com

7. *A light year is approximately 9.5×10^{12} km. If a star is 4 light years away, how far is it in kilometers?*

(A) 3.8×10^{13} *km*
(B) 3.8×10^{12} *km*
(C) 13.5×10^{12} *km*
(D) 9.5×10^{48} *km*

8. *A graph shows a straight line through $(0,0)$ and $(2,9)$. What is the unit rate?*

(A) 2
(B) 4.5
(C) 9
(D) 18

9. *A line passes through $(2,7)$ and $(5,1)$. What is the slope?*

(A) -2
(B) 2
(C) -3
(D) 3

10. *Solve $\frac{x}{4} + 3 = 7$.*

(A) $x = 1$
(B) $x = 10$
(C) $x = 16$
(D) $x = 28$

11. *Solve: $y = 4x$ and $y = -x + 15$.*

(A) $(2,8)$
(B) $(3,12)$
(C) $(5,10)$
(D) $(15,0)$

12. *The graph shows cost vs. number of months for two gym plans.*

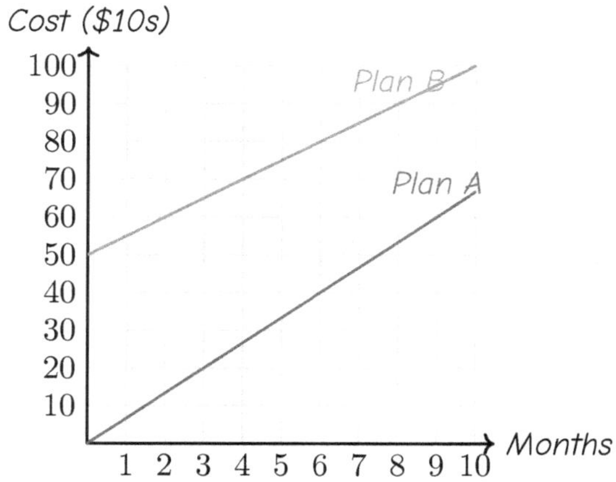

Plan A charges $\$20$ *per month (no fee). Plan B charges* $\$50$ *up front plus* $\$10$ *per month. After how many months does Plan A become more expensive?*

(A) 3 *months*

(B) 4 *months*

(C) 5 *months*

(D) 6 *months*

13. *The ordered pairs* $(a, 5)$ *and* $(a, 12)$ *belong to a relation. Can this relation be a function? Write Yes or No.*

Your Answer

14. *If* $f(x) = 5x + 2$, *which input gives an output of* 27?

(A) 3

(B) 4

(C) 5

(D) 6

15. *Function R is shown in the table:*

x	0	2	4	6
y	1	9	17	25

Function S: $y = 3x + 1$. Which has a greater initial value?

(A) *Function R*

(B) *Function S*

(C) *They have the same initial value.*

(D) *Cannot be determined.*

16. *Which of these is NOT a linear function?*

(A) $y = 0$

(B) $y = 100x$

(C) $y = -x + 5$

(D) $y = \sqrt{x}$

17. *A line passes through $(0, 4)$ and $(5, 24)$. Write its equation.*

Your Answer:

18. *A graph shows a line going upward with a constant slope. Is the rate of change constant or varying?*

Your Answer:

19. *A figure is reflected over the y-axis. What happens to point $(7, -2)$?*

(A) $(7, 2)$

(B) $(-7, 2)$

(C) $(-7, -2)$

(D) $(-2, 7)$

20. $\triangle ABC \cong \triangle DEF$. The perimeter of $\triangle ABC$ is 30 cm. What is the perimeter of $\triangle DEF$?

Your Answer

21. Point (a, b) is rotated $180°$ and then reflected over the x-axis. The final image is:

(A) (a, b)
(B) $(-a, b)$
(C) $(a, -b)$
(D) $(-a, -b)$

22. A map has a scale factor of $\frac{1}{50{,}000}$. If a road is 3 cm on the map, how long is the real road in meters?

Your Answer

23. Corresponding angles formed by a transversal and two parallel lines are located:

(A) On opposite sides of the transversal, between the parallel lines
(B) On the same side of the transversal, in matching positions
(C) On opposite sides of the transversal, outside the parallel lines
(D) Adjacent to each other at one intersection

24. A right triangle has legs 9 and 12. What is the hypotenuse?

(A) 21
(B) 108
(C) 15
(D) $\sqrt{21}$

25. What is the distance between $(-4, -3)$ and $(4, 3)$?

(A) $\sqrt{14}$
(B) 14
(C) 10
(D) $\sqrt{10}$

26. What is the volume of a cylinder with radius 2 cm and height 7 cm in terms of π?

(A) 14π cm^3
(B) 28π cm^3
(C) 56π cm^3
(D) 7π cm^3

27. A scatter plot of daily exercise (minutes) vs. resting heart rate shows a downward trend. How would you describe this?

(A) Positive association
(B) Negative association
(C) No association
(D) Clustering

28. A scatter plot has a clear downward trend. A student draws a horizontal line through the data. Is this a good fit?

(A) Yes — it splits the data evenly.
(B) No — it does not follow the downward trend.
(C) Yes — horizontal lines always fit data.
(D) No — the line should be vertical.

29. Using the model $y = -4x + 100$, predict y when $x = 10$.

(A) 140
(B) 60
(C) 96
(D) 40

30. What is a two-way frequency table?

(A) A table with two rows
(B) A table that shows frequencies for two categorical variables
(C) A table that shows only percentages
(D) A table with exactly two columns

Get Online
Find more at
ViewMath.com/RI-Grade8
ViewMath.com

End of Practice Test 6

Great job finishing the test!

My Score

I got ________ out of 30 questions right.

*Check your answers in the **Answer Key** at the back of the book.*

Review any questions you missed. That's how we learn!

Check Your Score Online!

*Visit **ViewMath Academy** to enter your answers and see which topics you need to review. You can also explore lessons, take quizzes, track your scores, and save your progress!*

viewmath.com/score/8.1.RI.21

*Or go to **viewmath.com/score** and enter code: **8.1.RI.21***

PRACTICE TEST

7

Practice Test 7

30 Questions

Before You Start

- **Read each question carefully** before choosing your answer.
- **Show your work** on scratch paper when you need to.
- **Skip hard questions** and come back to them later.
- **Check your answers** when you're done.
- **Take your time** — there's no rush!

You've Got This!

Do your best and show what you know!

1. *Give an example of an irrational number between* 1 *and* 2.

Your Answer:

2. *A student converts* $0.\overline{36}$ *and gets* $\frac{36}{100}$. *What mistake did the student make?*

(A) The student divided by 100 instead of 99.

(B) The student forgot to simplify.

(C) The student treated a repeating decimal as a terminating decimal.

(D) The student multiplied by 10 instead of 100.

3. *Between which two consecutive integers does* $\sqrt{83}$ *lie?*

Your Answer:

4. *A student claims that* $2\sqrt{5} = \sqrt{10}$. *Is this correct?*

(A) Yes, because $2 \times 5 = 10$.

(B) No, $2\sqrt{5} = \sqrt{20}$, not $\sqrt{10}$.

(C) Yes, multiplication distributes into the square root.

(D) No, $2\sqrt{5} = 4\sqrt{5}$.

5. *A cube-shaped box has a volume of* 216 *cubic centimeters. What is the edge length of the box?*

Your Answer:

6. *Which of the following is* $350{,}000$ *written in scientific notation?*

(A) 35×10^4

(B) 3.5×10^5

(C) 3.5×10^4

(D) 0.35×10^6

7. Earth's mass is about 6×10^{24} kg and Jupiter's mass is about 1.9×10^{27} kg. About how many times more massive is Jupiter than Earth? Round to the nearest whole number.

Your Answer

8. Which table does NOT represent a proportional relationship?

(A)

x	1	2
y	4	8

(B)

x	2	4
y	5	10

(C)

x	3	6
y	9	15

(D)

x	5	10
y	15	30

9. What is the slope of the line through $(1, 3)$ and $(4, 12)$?

(A) 2

(B) 3

(C) 4

(D) 9

10. Solve $0.5x + 1.5 = 4$.

(A) $x = 3$

(B) $x = 5$

(C) $x = 7$

(D) $x = 11$

11. Solve: $2x + 5y = 20$ and $2x + y = 8$.

Your Answer

12. *A snack bar sells hot dogs for $\$3$ and burgers for $\$5$. One day 80 items were sold for $\$340$. How many ho dogs were sold?*

(A) 20

(B) 30

(C) 40

(D) 50

13. *A vending machine gives one item for each button pressed. A student presses button A and sometimes get chips; other times the same button gives pretzels. Is this vending machine a function?*

(A) *Yes, because it always gives a snack.*

(B) *Yes, because button A is always pressed.*

(C) *No, because the same input (button A) gives two different outputs.*

(D) *No, because vending machines are not mathematical.*

14. *A function p is defined by the table.*

x	-2	-1	0	1	2
$p(x)$	9	4	1	0	1

For which input(s) is $p(x) = 1$?

(A) $x = 0$ *only*

(B) $x = 2$ *only*

(C) $x = 0$ *and* $x = 2$

(D) $x = -1$ *and* $x = 1$

15. *Function A: $y = 6x + 4$. Function B passes through $(0, 4)$ and $(3, 19)$. What is the rate of change of Function B?*

Your Answer:

16. *Which is a characteristic of a linear function?*

(A) Its graph is a curve.
(B) Its rate of change is constant.
(C) Its rate of change varies.
(D) It always passes through the origin.

17. *A taxi charges a $\$4$ flat fee plus $\$3$ per mile. Which equation models the total cost y for x miles?*

(A) $y = 4x + 3$
(B) $y = 3x + 4$
(C) $y = 7x$
(D) $y = 3x - 4$

18. *A decreasing graph does NOT necessarily mean:*

(A) The output values are getting smaller.
(B) The graph goes downward from left to right.
(C) The output values are negative.
(D) The rate of change is negative.

19. *A square is rotated $180°$ around its center. Which statement is true?*

(A) The square changes shape.
(B) The square changes size.
(C) The square maps onto itself.
(D) The square becomes a rectangle.

20. *Describe a sequence of rigid transformations that maps a triangle with vertices $(1, 1)$, $(4, 1)$, $(1, 3)$ to a triangle with vertices $(-1, 1)$, $(-4, 1)$, $(-1, 3)$.*

Your Answer:

21. *Which transformation maps (x, y) to $(-y, x)$?*

(A) Reflection over the x-axis
(B) Reflection over the y-axis
(C) $90°$ counterclockwise rotation
(D) $180°$ rotation

22. *All circles are similar to each other. Why?*

(A) They all have the same radius.

(B) One can always be dilated to match the other.

(C) They all have the same area.

(D) Circles cannot be dilated.

23. *A triangle has angles 62° and 73°. Find the third angle.*

Your Answer:

24. *A ladder leans against a wall as shown. How long is the ladder?*

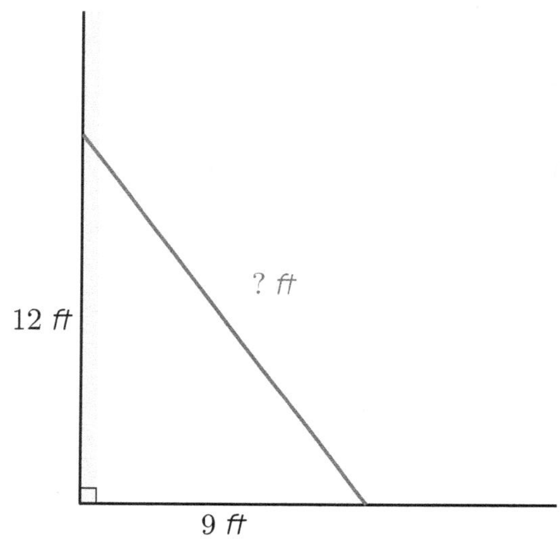

(A) 21 ft

(B) 3 ft

(C) 15 ft

(D) $\sqrt{21}$ ft

25. *Find the distance between $(5, -2)$ and $(5, 7)$.*

Your Answer:

26. The volume of a sphere with radius r is $\frac{4}{3}\pi r^3$. If the radius is tripled, the new volume is:

(A) 3 times the original

(B) 9 times the original

(C) 27 times the original

(D) 81 times the original

27. For the data: $(1, 10), (2, 8), (3, 7), (4, 5), (5, 3), (6, 2)$, plot the points and describe the association.

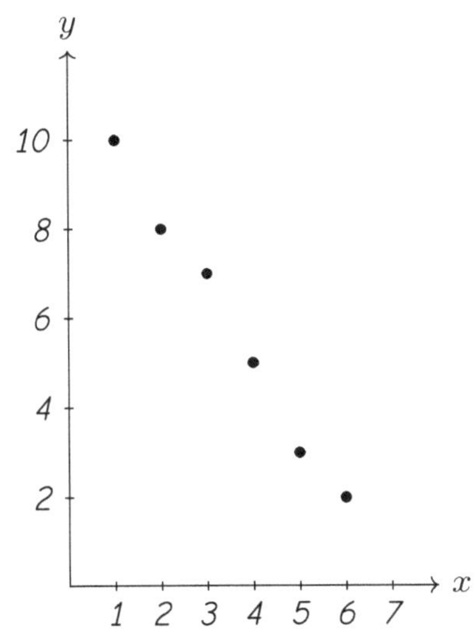

Your Answer:

28. What is a line of best fit?

(A) A line that passes through every data point

(B) A line that connects the first and last data points

(C) A straight line that comes close to most of the data points

(D) A vertical line through the middle of the scatter plot

29. *In the model* $y = 2.5x + 1$, *the actual value at* $x = 4$ *is* $y = 12$. *What is the residual?*

(A) 0

(B) 1

(C) -1

(D) 11

30. *If there is NO association between two variables in a two-way table, what would you expect?*

(A) *All joint frequencies equal zero.*

(B) *Relative frequencies are about the same across rows.*

(C) *One row has all the data.*

(D) *Marginal frequencies are always equal.*

★ End of Practice Test 7 ★

Great job finishing the test!

My Score

I got __________ out of 30 questions right.

*Check your answers in the **Answer Key** at the back of the book*

Review any questions you missed. That's how we learn!

Check Your Score Online!

*Visit **ViewMath Academy** to enter your answers and see which topics you need to review. You can also explore lessons, take quizzes, track your scores, and save your progress!*

viewmath.com/score/8.1.RI.22

*Or go to **viewmath.com/score** and enter code: 8.1.RI.22*

PRACTICE TEST

8

Practice Test 8

30 Questions

Before You Start

- **Read each question carefully** before choosing your answer.
- **Show your work** on scratch paper when you need to.
- **Skip hard questions** and come back to them later.
- **Check your answers** when you're done.
- **Take your time** — there's no rush!

You've Got This!

Do your best and show what you know!

1. Which of the following is a true statement?

(A) Every square root is irrational.

(B) Every integer is irrational.

(C) Every integer is rational.

(D) Every decimal is irrational.

2. The table below shows repeating decimals and their fraction equivalents. Find the missing fraction for $0.\overline{81}$.

Decimal	Unsimplified Fraction	Simplest Form
$0.\overline{27}$	$\frac{27}{99}$	$\frac{3}{11}$
$0.\overline{54}$	$\frac{54}{99}$	$\frac{6}{11}$
$0.\overline{81}$	?	?

Your Answer

3. Approximate $\sqrt{45}$ to one decimal place.

Your Answer

4. Which expression equals exactly 6?

(A) $\sqrt{3} \times \sqrt{12}$

(B) $\sqrt{3} + \sqrt{12}$

(C) $\sqrt{6} \times \sqrt{6}$

(D) Both A and C

5. Evaluate $\sqrt[3]{-125}$.

Your Answer

6. The bar chart compares the masses of four animals. Which animal's mass is closest to 1×10^2 kilograms?

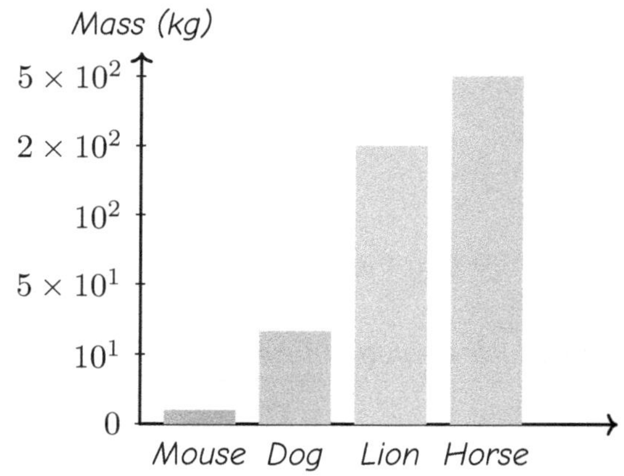

(A) Mouse

(B) Dog

(C) Lion

(D) Horse

7. A phone stores 6.4×10^{10} bytes. A photo uses 4×10^6 bytes. How many photos can the phone store?

(A) 1.6×10^3

(B) 1.6×10^4

(C) 1.6×10^{16}

(D) 2.4×10^4

8. Machine A fills $y = 12x$ bottles per hour. Machine B fills 50 bottles in 5 hours. Which machine is faster?

(A) Machine A

(B) Machine B

(C) They fill at the same rate.

(D) Cannot be determined.

9. The temperature drops from 68°F to 50°F over 6 hours. What is the rate of change in temperature?

(A) 3°F per hour

(B) -3°F per hour

(C) 6°F per hour

(D) -6°F per hour

10. How many solutions does $4x + 7 = 4x - 3$ have?

Your Answer

11. Look at the two lines graphed below. How many solutions does this system have?

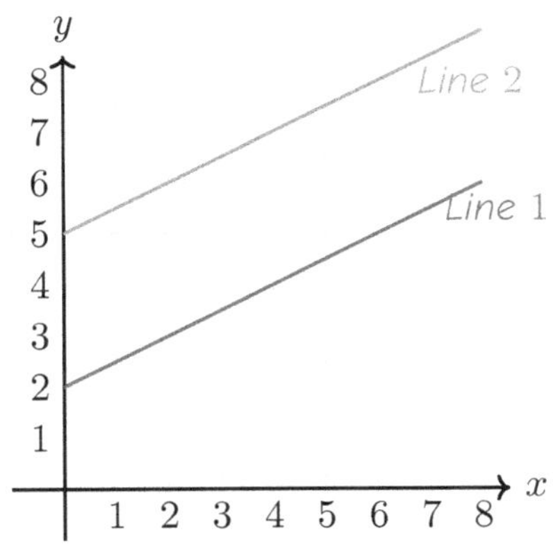

(A) One solution

(B) Two solutions

(C) No solution

(D) Infinitely many solutions

12. A farmer has chickens and cows. There are 20 heads and 56 legs. How many cows are there?

(A) 6

(B) 8

(C) 10

(D) 12

13. Does $y = x^2$ define a function? Write Yes or No.

Your Answer

Get Online
Find more at
ViewMath.com/RI-Grade8
ViewMath.com

14. The graph of h is shown below. For what value of x does $h(x) = 6$?

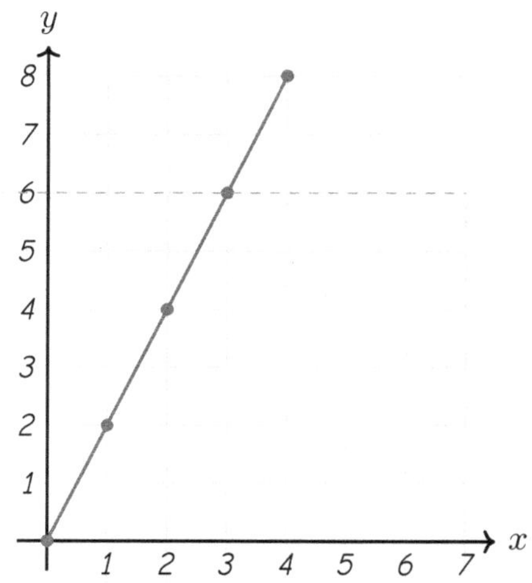

Your Answer:

15. A table for Function C shows $(0, 7)$, $(1, 11)$, $(2, 15)$. What is the initial value?

Your Answer:

16. A function increases by 6 for every increase of 1 in x. Is it linear or nonlinear?

Your Answer:

17. A bike rental costs $\$10$ plus $\$5$ per hour. What is the total cost for 3 hours?

Your Answer:

18. A car accelerates from a stop, cruises at constant speed, then brakes. Which describes the speed graph?

(A) Constant, increasing, decreasing

(B) Increasing, constant, decreasing

(C) Increasing, decreasing, constant

(D) Decreasing, constant, increasing

19. Name the three rigid transformations.

Your Answer

20. The two triangles shown below have matching angle measures. Are they congruent?

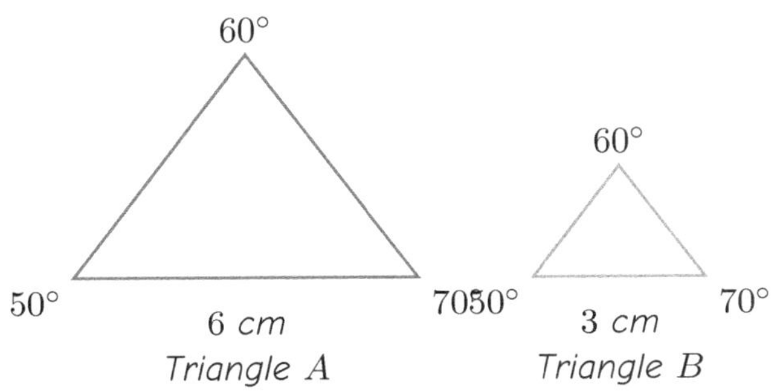

(A) Yes, because all three angles are equal.

(B) Yes, because one is a rotation of the other.

(C) No, because their side lengths are different.

(D) No, because their angles are in different positions.

21. What is the image of $(0, -8)$ after a $90°$ counterclockwise rotation around the origin?

Your Answer

Get Online
Find more at
ViewMath.com/RI-Grade8
ViewMath.com

22. *Triangle A has sides $4, 6, 8$. Triangle B has sides $6, 9, 12$. Are the triangles similar?*

(A) Yes, with scale factor $\frac{3}{2}$

(B) Yes, with scale factor 2

(C) No, the sides are not proportional.

(D) Yes, with scale factor 3

23. *An exterior angle of a triangle measures $125°$. The two non-adjacent interior angles are $60°$ and $x°$. What is x?*

(A) $55°$

(B) $65°$

(C) $115°$

(D) $185°$

24. *Is a triangle with sides $11, 60, 61$ a right triangle?*

Your Answer:

25. *Find the distance between $(-6, -8)$ and $(0, 0)$.*

Your Answer:

26. *What is the volume of a cylinder with radius 7 m and height 3 m? Leave your answer in terms of π.*

(A) $21\pi\ m^3$

(B) $63\pi\ m^3$

(C) $147\pi\ m^3$

(D) $49\pi\ m^3$

27. *Can a scatter plot show a positive association that is nonlinear? Give an example.*

Your Answer:

28. *A line passes through $(0, 3)$ and $(4, 11)$. What is the y-intercept?*

(A) 0

(B) 4

(C) 3

(D) 11

29. *A trend line gives $y = -3x + 45$. Data was collected for $x = 1$ to $x = 10$. Predict y when $x = 5$ and when $x = 30$. State which prediction is more reliable and why.*

Your Answer:

30. *Using the table above, what percentage of 7th graders ride a bike?*

(A) 55%

(B) 45%

(C) 62.5%

(D) 37.5%

★ End of Practice Test 8 ★

Great job finishing the test!

My Score

I got __________ out of 30 questions right.

*Check your answers in the **Answer Key** at the back of the book.*

Review any questions you missed. That's how we learn!

Check Your Score Online!

*Visit **ViewMath Academy** to enter your answers and see which topics you need to review. You can also explore lessons, take quizzes, track your scores, and save your progress!*

viewmath.com/score/8.1.RI.23

*Or go to **viewmath.com/score** and enter code: **8.1.RI.23***

PRACTICE TEST

9

Practice Test 9

30 Questions

Before You Start

- **Read each question carefully** before choosing your answer.
- **Show your work** on scratch paper when you need to.
- **Skip hard questions** and come back to them later.
- **Check your answers** when you're done.
- **Take your time** — there's no rush!

You've Got This!

Do your best and show what you know!

1. Which of the following could NOT be the decimal expansion of a rational number?

(A) 2.500
(B) $0.\overline{9}$
(C) $1.41421356\ldots$ (non-repeating)
(D) $0.\overline{36}$

2. What is $0.\overline{4}$ written as a fraction?

(A) $\frac{4}{10}$
(B) $\frac{4}{9}$
(C) $\frac{2}{5}$
(D) $\frac{1}{4}$

3. A student says $\sqrt{45}$ is between 6 and 7. Is the student correct?

(A) Yes, because $6^2 = 36$ and $7^2 = 49$.
(B) No, it is between 5 and 6.
(C) No, it is between 7 and 8.
(D) No, $\sqrt{45} = 9$ exactly.

4. Which is larger: $\sqrt{2} + \sqrt{3}$ or $\sqrt{10}$?

(A) $\sqrt{2} + \sqrt{3}$, because $2 + 3 = 5 > \sqrt{10}$
(B) $\sqrt{10}$, because $10 > 5$
(C) $\sqrt{10}$, because $\sqrt{10} \approx 3.16 > 3.15 \approx \sqrt{2} + \sqrt{3}$
(D) They are equal.

5. Between which two consecutive whole numbers does $\sqrt{50}$ lie?

(A) 6 and 7
(B) 7 and 8
(C) 24 and 26
(D) 8 and 9

6. Which is larger: 8×10^5 or 3×10^6?

(A) 8×10^5, because $8 > 3$
(B) 3×10^6, because the exponent is larger
(C) They are equal
(D) It cannot be determined

Get Online
Find more at
ViewMath.com/RI-Grade8
ViewMath.com

7. *Simplify* $(7 \times 10^{-3})(8 \times 10^{-2})$. *Write in proper scientific notation.*

(A) 56×10^{-5}

(B) 5.6×10^{-5}

(C) 5.6×10^{-4}

(D) 5.6×10^{6}

8. *Two delivery services are compared. Service P delivers* $y = 4x$ *packages per hour. Service Q's data is shown in the table.*

Hours (x)	2	4	6	8
Packages (y)	10	20	30	40

How many more packages does Service Q deliver than Service P in 10 *hours?*

Your Answer

9. *Find the slope through* $(0, -2)$ *and* $(3, 7)$.

(A) 3

(B) -3

(C) $\frac{5}{3}$

(D) $\frac{9}{3}$

10. *The diagram shows tiles representing an equation. Solve for* x.

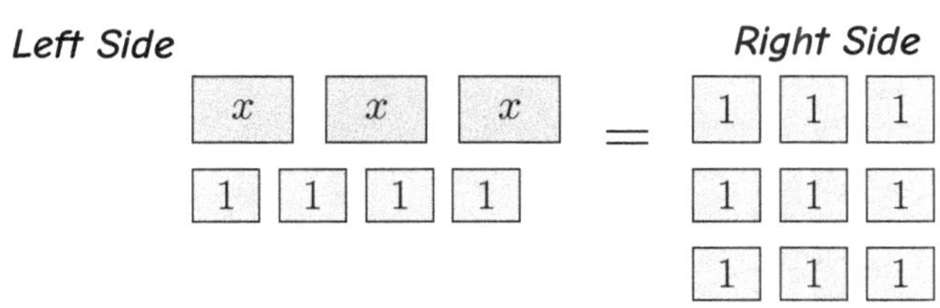

Your Answer

11. *Solve using elimination:* $3x + 2y = 16$ *and* $x + 2y = 8$.

(A) $(2, 5)$
(B) $(4, 2)$
(C) $(3, 3)$
(D) $(6, -1)$

12. *Adult tickets cost* $\$10$ *and child tickets cost* $\$6$. *A family buys* 5 *tickets for* $\$38$ *total. How many adult tickets did they buy?*

(A) 1
(B) 2
(C) 3
(D) 4

13. *Which of the following tables does NOT represent a function?*

x	2	3	4	3	5
y	6	9	12	15	18

(A) *It is a function because all* y*-values are different.*
(B) *It is NOT a function because input* 3 *gives two different outputs* (9 *and* 15)*.*
(C) *It is a function because the* y*-values increase.*
(D) *It is NOT a function because* $x = 2$ *is too small.*

14. *If* $g(x) = x^2 - 5$, *what is* $g(3)$*?*

(A) -2
(B) 1
(C) 4
(D) 14

15. *Plan X costs* $\$40$ *plus* $\$5$ *per month. Plan Y costs* $\$10$ *plus* $\$10$ *per month. After how many months will they cost the same?*

(A) 4 *months*
(B) 6 *months*
(C) 8 *months*
(D) 10 *months*

Get Online

Find more at
ViewMath.com/RI-Grade8

ViewMath.com

16. The function $y = -2x + 6$ is:

(A) Nonlinear, because the slope is negative.

(B) Nonlinear, because y can be negative.

(C) Linear, because it has the form $y = mx + b$.

(D) Linear, because it passes through the origin.

17. A line passes through $(2, 10)$ and $(5, 22)$. What is the equation?

(A) $y = 4x + 2$

(B) $y = 3x + 4$

(C) $y = 4x + 10$

(D) $y = 12x - 2$

18. A student walks to a friend's house, waits for the friend, then they walk together to school. What does the distance-from-home graph look like?

(A) Increasing, constant, increasing

(B) Increasing, decreasing, increasing

(C) Constant, increasing, constant

(D) Decreasing, constant, decreasing

19. Which of the following is a rigid transformation?

(A) Dilation by a factor of 2

(B) Stretching a figure horizontally

(C) Reflection over a line

(D) Shrinking a figure to half its size

20. How many rigid transformations are needed to map Figure A onto Figure B if B is A reflected over the x-axis and then translated 3 units right?

(A) 1

(B) 2

(C) 3

(D) 0

Get Online

Find more at
ViewMath.com/RI-Grade8

ViewMath.com

21. *A triangle with vertices* $(0,0)$, $(6,0)$, *and* $(6,8)$ *is dilated by factor* $\frac{1}{2}$ *from the origin. What is the perimeter of the image?*

(A) 24

(B) 12

(C) 6

(D) 48

22. *Two figures are similar if one can be mapped onto the other by a sequence of:*

(A) *Only translations*

(B) *Only dilations*

(C) *Rigid transformations and dilations*

(D) *Only reflections*

23. *Two parallel lines are cut by a transversal. Angle* $1 = (3x+10)^\circ$ *and Angle* $2 = (5x-30)^\circ$ *are alternate interior angles. What is* x*?*

(A) 10

(B) 20

(C) 25

(D) 40

24. *Which equation represents the Pythagorean Theorem?*

(A) $a + b = c$

(B) $a^2 + b^2 = c^2$

(C) $a^2 \times b^2 = c^2$

(D) $2a + 2b = c$

Get Online

Find more at
ViewMath.com/RI-Grade8

ViewMath.com

25. *Two points are shown on the coordinate plane. What is the distance between them?*

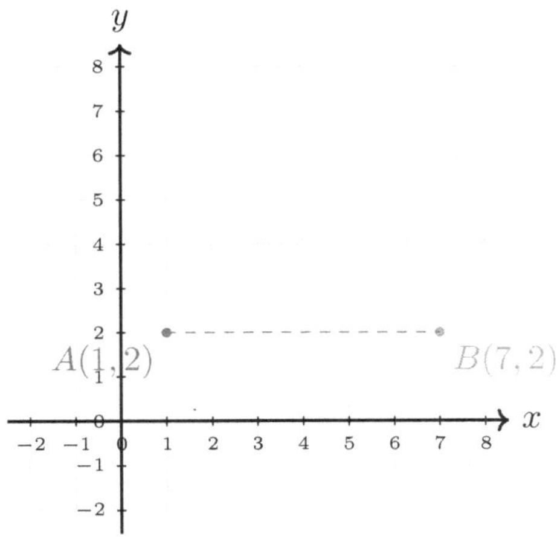

(A) 8

(B) 6

(C) 4

(D) $\sqrt{6}$

26. *A cylinder has volume* 200π *cm*^{3} *and height* 8 *cm. What is the radius?*

(A) 25 *cm*

(B) 5 *cm*

(C) $\sqrt{25}$ *cm*

(D) 10 *cm*

27. *Identify the outlier in the scatter plot.*

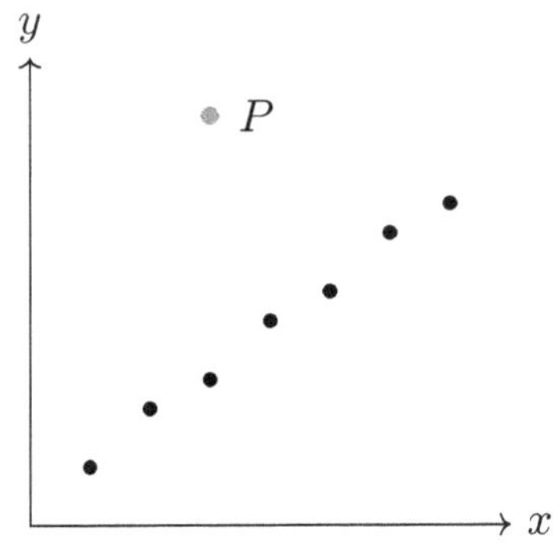

(A) $(1,1)$

(B) $(7,5.5)$

(C) $(3,7)$ — *point* P

(D) $(4,3.5)$

28. *A trend line has equation* $y=-3x+25$. *What is the slope?*

(A) 25

(B) 3

(C) -3

(D) -25

29. *A positive residual means:*

(A) *The predicted value is higher than the actual value.*

(B) *The actual value is higher than the predicted value.*

(C) *The slope of the model is positive.*

(D) *The data point is an outlier.*

30.

	Left hand	Right hand	Total
Male	?	35	40
Female	8	?	40
Total	?	?	80

Fill in the missing values.

Your Answer

End of Practice Test 9

Great job finishing the test!

My Score

I got __________ out of 30 questions right.

Check your answers in the ***Answer Key*** *at the back of the book.*

Review any questions you missed. That's how we learn!

Check Your Score Online!

Visit ***ViewMath Academy*** *to enter your answers and see which topics you need to review. You can also explore lessons, take quizzes, track your scores, and save your progress!*

viewmath.com/score/8.1.RI.24

Or go to ***viewmath.com/score*** *and enter code:* ***8.1.RI.24***

PRACTICE TEST

10

Practice Test 10

30 Questions

Before You Start

- **Read each question carefully** before choosing your answer.
- **Show your work** on scratch paper when you need to.
- **Skip hard questions** and come back to them later.
- **Check your answers** when you're done.
- **Take your time** — there's no rush!

You've Got This!

Do your best and show what you know!

1. *Look at the Venn diagram below. In which region does $\sqrt{5}$ belong?*

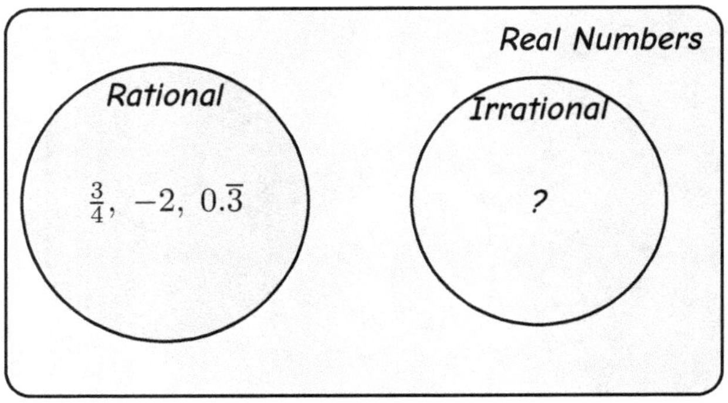

Ⓐ *Rational region, because* $\sqrt{5} = 2.5$

Ⓑ *Irrational region, because* 5 *is not a perfect square*

Ⓒ *Rational region, because* $\sqrt{5}$ *is a square root*

Ⓓ *Neither region, because* $\sqrt{5}$ *is not a real number*

2. *The steps below show a student's work for converting* $0.\overline{2}$ *to a fraction. Which step contains the error?*

Step	Work
1	Let $x = 0.222\ldots$
2	$10x = 2.222\ldots$
3	$10x - x = 2.222\ldots - 0.222\ldots$
4	$9x = 2$
5	$x = \frac{2}{10}$

Ⓐ *Step 2*

Ⓑ *Step 3*

Ⓒ *Step 4*

Ⓓ *Step 5*

3. *Between which two consecutive integers does* $\sqrt{50}$ *lie?*

Ⓐ 6 *and* 7

Ⓑ 7 *and* 8

Ⓒ 8 *and* 9

Ⓓ 24 *and* 26

4. Order from least to greatest: π, $\sqrt{10}$, 3.2.

Ⓐ $\pi,\ 3.2,\ \sqrt{10}$

Ⓑ $3.2,\ \pi,\ \sqrt{10}$

Ⓒ $\sqrt{10},\ \pi,\ 3.2$

Ⓓ $\pi,\ \sqrt{10},\ 3.2$

5. Which statement about $\sqrt{2}$ is true?

Ⓐ $\sqrt{2}$ is a rational number

Ⓑ $\sqrt{2}$ is an integer

Ⓒ $\sqrt{2}$ is an irrational number

Ⓓ $\sqrt{2} = 1.5$

6. The diagram below shows three cells viewed under a microscope, with their actual sizes labeled. Write all three sizes in scientific notation, then list them from smallest to largest.

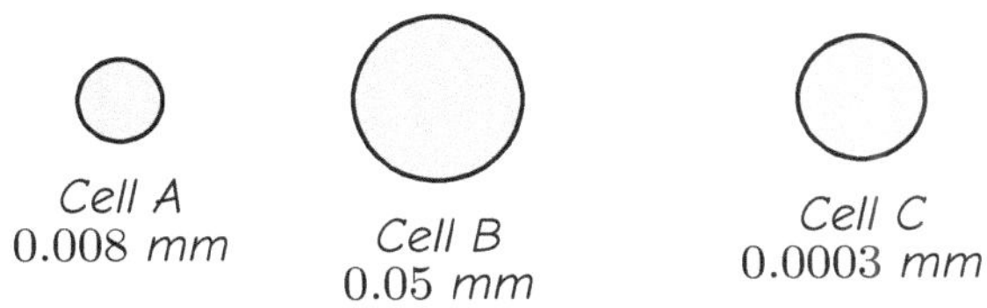

Your Answer:

7. Compute $\frac{3.6\times10^{10}}{1.2\times10^{4}}$. Write your answer in scientific notation.

Your Answer:

8. If $y = kx$ and $y = 54$ when $x = 6$, find y when $x = 11$.

Your Answer:

9. *A line goes down from left to right. Its slope must be:*

(A) Positive
(B) Negative
(C) Zero
(D) Undefined

10. *How many solutions does $6x + 4 = 6x + 4$ have?*

(A) No solution
(B) One solution
(C) Two solutions
(D) Infinitely many solutions

11. *Use the tables to find the solution to the system of equations.*

Equation 1: $y = 2x + 1$

x	y
0	1
1	3
2	5
3	7
4	9

Equation 2: $y = -x + 7$

x	y
0	7
1	6
2	5
3	4
4	3

Your Answer

12. *A rectangle's perimeter is 34 cm. The length is 5 cm more than the width. What is the width?*

(A) 4 cm
(B) 5 cm
(C) 6 cm
(D) 7 cm

13. *Use the mapping diagram below. List all the inputs that are part of this relation.*

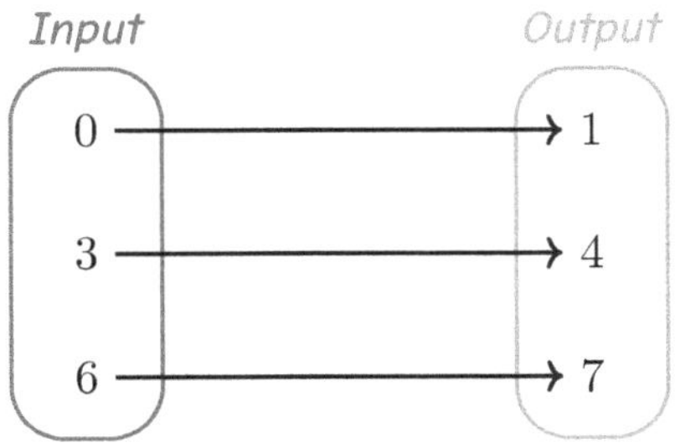

Your Answer

14. *If* $f(x) = 7 - x$*, what is* $f(10)$*?*

Ⓐ -3

Ⓑ 3

Ⓒ 17

Ⓓ -17

15. *Function G is shown in the table:*

x	0	1	2	3	4
y	2	5	8	11	14

Function H: $y = 4x + 1$*. Which function has the greater value at* $x = 4$*?*

Ⓐ *Function G*

Ⓑ *Function H*

Ⓒ *They are equal.*

Ⓓ *Cannot be determined.*

16. *Is* $y = 8x + 1$ *linear or nonlinear?*

Your Answer

17. *A line passes through $(3,7)$ and $(6,16)$. What is the y-intercept?*

(A) -2

(B) 0

(C) 1

(D) 3

18. *Describe the behavior of the function shown in the graph below. Identify each section as increasing, decreasing, or constant, and whether it is linear or nonlinear.*

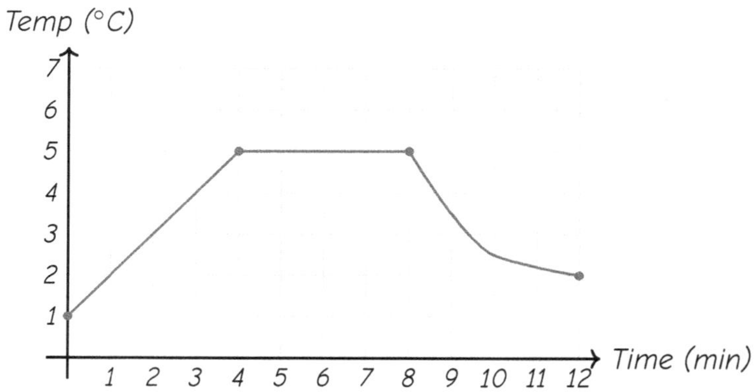

Your Answer:

19. *After a translation, a triangle's longest side was originally 8 cm. What is the longest side of the image?*

(A) 4 cm

(B) 8 cm

(C) 16 cm

(D) It depends on the direction of the translation.

20. *Congruent figures must have:*

(A) The same position on the plane

(B) The same color and orientation

(C) The same side lengths and angle measures

(D) The same area but different perimeters

21. *What is the image of $(5, -3)$ after a translation of $(-2, 4)$?*

Ⓐ $(7, -7)$

Ⓑ $(3, 1)$

Ⓒ $(3, -7)$

Ⓓ $(7, 1)$

22. *Are all squares similar to each other?*

Ⓐ *No, because they can have different side lengths.*

Ⓑ *No, because they can have different angles.*

Ⓒ *Yes, because all squares have equal angles and proportional sides.*

Ⓓ *Yes, but only if they have the same perimeter.*

23. *A co-interior angle measures $97°$. What is the other co-interior angle?*

Your Answer:

24. *Is the triangle with the given side lengths a right triangle?*

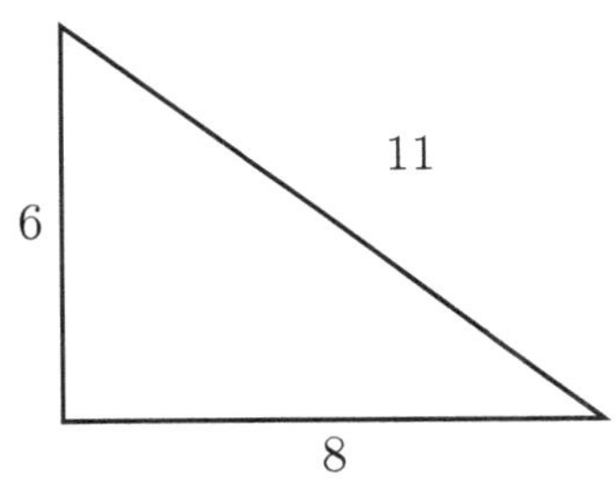

Ⓐ *Yes, because* $6^2 + 8^2 = 11^2$

Ⓑ *Yes, because* $6 + 8 > 11$

Ⓒ *No, because* $6^2 + 8^2 \neq 11^2$

Ⓓ *No, because the triangle has no right angle symbol*

Get Online
Find more at
ViewMath.com/RI-Grade8
ViewMath.com

25. *What is the distance between $(0, -6)$ and $(8, 0)$?*

(A) 14

(B) $\sqrt{14}$

(C) 10

(D) 2

26. *A cylindrical can has radius 3 in and height 8 in. How much soup can it hold? Use $\pi \approx 3.14$*

Your Answer

27. *A scatter plot of car age (x) versus car value (y) shows dots going from upper-left to lower-right. What type of association is this?*

(A) *Positive association*

(B) *Negative association*

(C) *No association*

(D) *Constant association*

28. *Which line is the best fit for the data shown?*

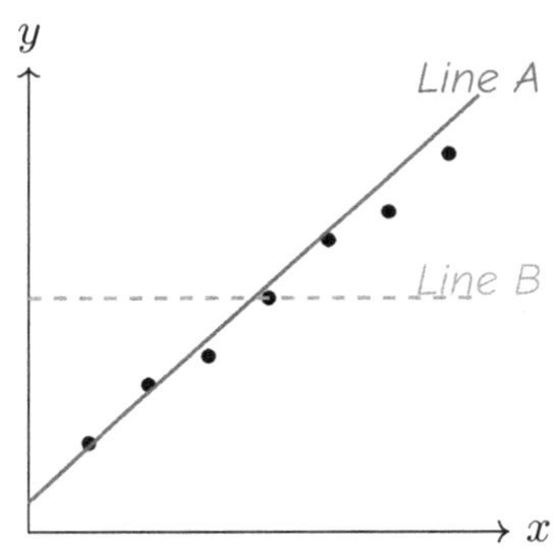

(A) *Line A — it follows the upward trend*

(B) *Line B — it splits the data horizontally*

(C) *Neither — no line fits this data*

(D) *Both lines fit equally well*

29. A model predicts that a car depreciates according to $y = -1{,}500x + 20{,}000$ where x is years and y is value. What is the car's predicted value after 5 years?

(A) $\$12{,}500$

(B) $\$15{,}000$

(C) $\$7{,}500$

(D) $\$27{,}500$

30. Based on the table above, is there an association between studying and passing?

(A) No — the pass rates are the same.

(B) Yes — students who studied had a higher pass rate.

(C) No — more students did not study.

(D) Yes — but only for students who failed.

End of Practice Test 10

Great job finishing the test!

My Score

I got __________ out of 30 questions right.

*Check your answers in the **Answer Key** at the back of the book.*

Review any questions you missed. That's how we learn!

Check Your Score Online!

*Visit **ViewMath Academy** to enter your answers and see which topics you need to review. You can also explore lessons, take quizzes, track your scores, and save your progress!*

viewmath.com/score/8.1.RI.25

*Or go to **viewmath.com/score** and enter code: **8.1.RI.25***

Answer Key & Explanations

Answer Key

First try each test on your own, then check your work here.

Practice Test 1 — Answer Key

1. False
2. C
3. D
4. B
5. B
6. $\approx 3.1 \times 10^{12}$
7. C
8. B
9. 3
10. $x = 13$
11. B
12. 3 hours; 150 miles from home
13. 1
14. C
15. 6 hours
16.
17. $y = 3x$
18. The speed is constant for 4 seconds.
19. C
20. C
21. B
22. D
23. B
24. B
25. C
26. B
27. C
28. Slope $= -1$; y-intercept $= 12$
29. B
30. A joint frequency is the count for a specific combination of two categories. A marginal frequency is the total o

Time to Learn!

*Review the explanations below, **especially for the questions you missed**.*

Understanding why each answer is correct builds stronger problem-solving skills.

Tip: Circle any questions you got wrong, then read their explanation carefully.

Practice Test 1 — Detailed Explanations

1 $\frac{22}{7} \approx 3.142857\ldots$ *is a rational approximation of* π, *but* $\pi = 3.14159265\ldots$ *is irrational. They are close but not equal.*

2 $0.\overline{123}$ *has a 3-digit repeating block, so you should multiply by* $10^3 = 1000$, *not* 100.

3 $3.74^2 = 13.9876$ *(under) and* $3.75^2 = 14.0625$ *(just over).* 14.0625 *is very close to* 14, *so* 3.75 *is the best next estimate.*

4 $\sqrt{7} \approx 2.646$, *so* $5 - \sqrt{7} \approx 5 - 2.646 = 2.354 \approx 2.4$.

5 *If* $s = 8$, *then* $A = 8^2 = 64$. *If* $e = 4$, *then* $V = 4^3 = 64$. *Since* $A = V = 64$, *the pair* $s = 8, e = 4$ *works.*

6 $498{,}000{,}000 \approx 5 \times 10^8$ *and* $6{,}200 \approx 6.2 \times 10^3$. *Product* $\approx 5 \times 6.2 \times 10^{11} = 31 \times 10^{11} = 3.1 \times 10^{12}$.

7 $(3 \times 10^4)(5 \times 10^3) = 15 \times 10^7 = 1.5 \times 10^8$.

8 *At \$8 per hour with no extra fee, the relationship is proportional:* $y = 8x$.

9 $m = \frac{11-(-1)}{7-3} = \frac{12}{4} = 3$.

10 *Multiply by* 2: $x + 5 = 18$. *Subtract* 5: $x = 13$.

11 $2x = -x + 9$, *so* $3x = 9$, $x = 3$. *Then* $y = 2(3) = 6$. *Intersection:* $(3, 6)$.

12 *Car:* $d = 50t$. *Bus:* $d = 40t + 30$. *Set equal:* $50t = 40t + 30$, $10t = 30$, $t = 3$. *Distance:* $50(3) = 150$ *miles.*

13 *By the vertical line test, if a vertical line crosses a graph more than once, it is not a function. So a function's graph is crossed at most once.*

14 $f(4) = 3(4) + 1 = 12 + 1 = 13.$

15 $8x + 20 = 3x + 50 \Rightarrow 5x = 30 \Rightarrow x = 6.$

16 $y = x^2 - 1$ *has x raised to the 2nd power, making it nonlinear. The others are all in the form* $y = mx + b$.

17 *Slope:* $\frac{12-3}{4-1} = \frac{9}{3} = 3$. *Using* $(1, 3)$: $3 = 3(1) + b \Rightarrow b = 0$. *Equation:* $y = 3x$.

18 *A flat section means the output (speed) stays the same — the object travels at a steady speed for those 4 seconds.*

19 *A reflection reverses the orientation (like looking at a mirror image). Side lengths, angles, and area are all preserved.*

20 *By definition, congruent figures can be mapped onto each other using rigid transformations (translations, reflections, rotations).*

21 *Reflecting over the x-axis keeps the x-coordinate and negates the y-coordinate:* $(x, y) \to (x, -y)$.

22 *The longest side is* 13. *Multiplied by* $k = 2$: $13 \times 2 = 26$.

23 *Alternate interior angles formed by parallel lines and a transversal are equal:* $72°$.

24 $c^2 = 10^2 + 10^2 = 200$, *so* $c = \sqrt{200} = 10\sqrt{2}$.

25 $d = \sqrt{(2-(-1))^2 + (6-2)^2} = \sqrt{9 + 16} = \sqrt{25} = 5.$

26 $V = \frac{4}{3}(3.14)(125) = \frac{4}{3}(392.5) \approx 523.3$ *in*^{3}.

Get Online
Find more at
ViewMath.com/RI-Grade8
ViewMath.com

27 *When analyzing a scatter plot, you describe direction, shape, outliers, and clusters. Finding the exact equation requires additional calculation.*

28 $m = \frac{4-10}{8-2} = \frac{-6}{6} = -1$. $10 = -1(2) + b$, *so* $b = 12$.

29 $0 = -0.5x + 25$, $0.5x = 25$, $x = 50$.

30 *Joint = interior cell (intersection). Marginal = row/column total at the edges.*

Practice Test 2 — Answer Key

1 C

2 *Multiply by* 100; $100x = 54.\overline{54}$; $99x = 54$; $x = \frac{54}{99} = \frac{6}{11}$

3 A: ≈ 4.2, B: ≈ 6.3, C: ≈ 8.5

4 C

5 B

6 B

7 A

8 180 mL

9 B

10 Infinitely many

11 B

12 7.5 km/h

13 Yes, it is a function.

14 5

15 A

16 B

17 A

18 B

19 $(-4, 2)$

20 C

21 A

22 False

23 B

24 $a^2 + 16a^2 = 17a^2$

25 C

26 ≈ 111.3 cm^3

27 C

28 A

29 B

30 B

Time to Learn!

Review the explanations below, ***especially for the questions you missed.***

Understanding why each answer is correct builds stronger problem-solving skills.

Tip: *Circle any questions you got wrong, then read their explanation carefully.*

Get Online
Find more at
ViewMath.com/RI-Grade8
ViewMath.com

Practice Test 2 — Detailed Explanations

1 $P=\sqrt{2}$ is irrational (since 2 is not a perfect square) and $Q=3$ is an integer, which is rational.

2 The repeating block has 2 digits, so multiply by 100. Then $100x-x=99x=54$, giving $x=\frac{54}{99}=\frac{6}{11}$.

3 Side $=\sqrt{Area}$. $\sqrt{18}\approx 4.24\approx 4.2$. $\sqrt{40}\approx 6.32\approx 6.3$. $\sqrt{72}\approx 8.49\approx 8.5$.

4 $\sqrt{5}\approx 2.236$ and $\sqrt{3}\approx 1.732$. Adding: $2.236+1.732=3.968\approx 4.0$.

5 $x=\pm\sqrt{\frac{9}{16}}=\pm\frac{3}{4}$, since $\left(\frac{3}{4}\right)^2=\frac{9}{16}$.

6 $\frac{4.2\times10^3}{4.2\times10^{-3}}=10^{3-(-3)}=10^6$. So 4.2×10^3 is 10^6 (one million) times as large.

7 Divide the coefficients: $\frac{8}{2}=4$. Subtract the exponents: $10^{7-3}=10^4$. Answer: 4×10^4.

8 Rate $=\frac{45}{15}=3$ mL/min. In 60 minutes: $3\times60=180$ mL.

9 Rate of change $=\frac{16-4}{6}=\frac{12}{6}=2$ cm per week.

10 $2x+6=2x+6$ simplifies to $0=0$. Every value of x works.

11 Substitute $x=y+2$: $3(y+2)+y=14$. So $3y+6+y=14$, $4y=8$, $y=2$. Then $x=4$.

12 $b+c=9$ and $b-c=6$. Add: $2b=15$, $b=7.5$ km/h.

13 The graph is a straight line. Any vertical line will cross it at exactly one point, so it passes the vertical line test and is a function.

14 $-3x + 15 = 0 \Rightarrow -3x = -15 \Rightarrow x = 5.$

15 Function A starts at \$50 ($b = 50$). Function B starts at \$30 ($b = 30$). Since $50 > 30$, Function A starts higher.

16 $7 - 3 = 4$, $11 - 7 = 4$, $15 - 11 = 4$. The constant change of 4 proves this is linear.

17 The slope $m = 50$ represents the amount added per week, so the weekly deposit is \$50.

18 A steeper upward slope means faster increase. The morning section is steeper than the afternoon section, so the temperature rose faster in the morning.

19 Subtract 6 from x and add 3 to y: $(2 - 6, -1 + 3) = (-4, 2)$.

20 Two rectangles can have the same perimeter but different dimensions (e.g., 2×6 and 3×5 both have perimeter 16), so they are not necessarily congruent.

21 Apply the rule: $(-1, 6) \to (6, -(-1)) = (6, 1)$.

22 A 2×5 rectangle and a 3×4 rectangle have the same angles ($90°$) but their side ratios differ ($\frac{2}{5} \neq \frac{3}{4}$), so they are not similar.

23 Co-interior angles are supplementary: $180 - 135 = 45°$.

24 $a^2 + (4a)^2 = a^2 + 16a^2 = 17a^2 = (\sqrt{17} \cdot a)^2$. Confirmed.

25 The right triangle has legs 8 and 6. Distance $= \sqrt{8^2 + 6^2} = \sqrt{64 + 36} = \sqrt{100} = 10$.

26 Cone: $\frac{1}{3}(3.14)(6.25)(12) = \frac{1}{3}(235.5) = 78.5\ cm^3$. Half-sphere: $\frac{1}{2} \times \frac{4}{3}(3.14)(15.625) = \frac{1}{2}(65.42) \approx 32.7\ cm^3$. Total: $78.5 + 32.7 \approx 111.2\ cm^3$.

27 When both variables increase together, the association is positive.

28 $m = \frac{8-3}{5-1} = \frac{5}{4}$.

29 Residual = actual − predicted $= 8 - 6 = 2$.

30 60% of boys chose band vs. 45% of girls. The difference in conditional proportions shows an association.

Practice Test 3 — Answer Key

1 C	2 C	3 B	4 C	5 C	6 A	7 A	8 A	9 B	10 D
11 B	12 16 and 20	13 C	14 7	15 B	16 C	17 C	18 B	19 B	
20 C	21 B	22 10	23 B	24 B	25 C	26 300 cm^3	27 C	28 B	
29 B	30 A								

Time to Learn!

Review the explanations below, **especially for the questions you missed.**

Understanding why each answer is correct builds stronger problem-solving skills.

Tip: Circle any questions you got wrong, then read their explanation carefully.

Practice Test 3 — Detailed Explanations

1 $0.\overline{81}$ *is a repeating decimal, so it can be written as a fraction:* $0.\overline{81} = \frac{81}{99} = \frac{9}{11}$. *The others are irrational.*

2 *Let* $x = 0.\overline{27}$. *Then* $100x = 27.\overline{27}$. *Subtract:* $99x = 27$, *so* $x = \frac{27}{99} = \frac{3}{11}$.

3 $7.7^2 = 59.29$ *and* $7.8^2 = 60.84$. *So* $\sqrt{60} \approx 7.75$, *making* 8 *a less accurate estimate.* 7.7 *or* 7.8 *is better.*

4 $4\sqrt{2} \approx 4 \times 1.414 = 5.656$ *and* $\sqrt{30} \approx 5.477$. *Since* $5.656 > 5.477$, $4\sqrt{2}$ *is larger.*

5 *Side* $= \sqrt{196} = 14$ *cm, since* $14 \times 14 = 196$.

6 *Move the decimal 4 places right:* $0.00072 = 7.2 \times 10^{-4}$.

7 $(9.5 \times 10^{-18})(2 \times 10^{6}) = 19 \times 10^{-12} = 1.9 \times 10^{-11}$ *g.*

8 *Store A:* $k = 2$ *per pound. Store B:* $k = \frac{7.50}{3} = 2.50$ *per pound. Store A is cheaper.*

9 $\frac{270-120}{5-2} = \frac{150}{3} = 50$ *mph.*

10 $5x + 10 = 5x + 10$. *Subtract* $5x + 10$: $0 = 0$. *This is always true — infinitely many solutions.*

11 $\frac{1}{2}x + 3 = 2x - 3$. *Subtract* $\frac{1}{2}x$: $3 = \frac{3}{2}x - 3$. *Add 3:* $6 = \frac{3}{2}x$, *so* $x = 4$. *Then* $y = 2(4) - 3 = 5$.

12 $x = y + 4$ *and* $x + y = 36$. *Substitute:* $(y + 4) + y = 36$, $2y = 32$, $y = 16$. *Then* $x = 20$.

13 *Each input* $(1, 2, 3, 4)$ *appears once and maps to exactly one output. That is the definition of a function.*

14 $g(-2) = (-2)^2 + 3 = 4 + 3 = 7.$

15 *Set* $2x + 100 = 8x + 10$. *Subtract* $2x$: $100 = 6x + 10$. *Subtract* 10: $90 = 6x$. *Divide:* $x = 15$.

16 $6 - 3 = 3$, $11 - 6 = 5$, $18 - 11 = 7$. *The* y*-differences are not the same, so the rate of change is not constant — nonlinear.*

17 *Slope:* $10 - 7 = 3$. y*-intercept:* 7 *(when* $x = 0$*). Equation:* $y = 3x + 7$.

18 *Losing fuel at a steady rate means the amount decreases at a constant rate — that's a straight line going down (linear and decreasing).*

19 *The* $90°$ *CCW rule is* $(x, y) \rightarrow (-y, x)$. *So* $(-1, 4) \rightarrow (-4, -1)$.

20 *Applying the* $90°$ *CCW rule* $(x, y) \rightarrow (-y, x)$: $R(0, 0) \rightarrow (0, 0)$, $S(4, 0) \rightarrow (0, 4)$, $T(0, 3) \rightarrow (-3, 0)$. *This matches* $R'S'T'$.

21 *Divide the image coordinates by the scale factor:* $(9 \div 3,\ -15 \div 3) = (3, -5)$.

22 *The hypotenuse of Triangle* A *is* 5. *Multiply by the scale factor:* $5 \times 2 = 10$.

23 *Alternate interior angles are equal:* $4x = 80$, *so* $x = 20$.

24 $d^2 = 36^2 + 27^2 = 1296 + 729 = 2025$, *so* $d = \sqrt{2025} = 45$ *in.*

25 *The side length is* 6. *The diagonal of a square with side* s *is* $s\sqrt{2} = 6\sqrt{2} \approx 8.49$.

26 *A cylinder is* 3 *times the volume of a cone with the same base and height:* $100 \times 3 = 300$ *cm*^{3}.

27 A strong positive trend corresponds to an r-value close to 1. $r = 0.9$ indicates a strong positive association.

28 $m = \frac{12-4}{5-1} = \frac{8}{4} = 2$.

29 Growth per year is the rate of change, which is the slope of the linear model.

30 $\frac{24}{80} = 0.30$.

Practice Test 4 — Answer Key

1 D | 2 $\frac{53}{90}$ | 3 C | 4 6 | 5 B | 6 C | 7 C | 8 Taxi B | 9 C

10 C | 11 D | 12 5 | 13 B | 14 3 | 15 B | 16 B | 17 B | 18 B | 19 C

20 C | 21 B | 22 B | 23 B | 24 26 | 25 D | 26 C | 27 A | 28 A

29 Predicted height at day 10: $y = 0.4(10) + 2 = 6$ cm. Slope: the plant grows 0.4 cm per day. y-intercept: the plant starts

30 B

Time to Learn!

Review the explanations below, **especially for the questions you missed.**

Understanding why each answer is correct builds stronger problem-solving skills.

Tip: Circle any questions you got wrong, then read their explanation carefully.

Practice Test 4 — Detailed Explanations

1 $\sqrt{7}$ is irrational because 7 is not a perfect square. The other choices are all rational: $\frac{5}{8} = 0.625$, $0.\overline{6} = \frac{2}{3}$, and $\sqrt{9} = 3$.

2 Let $x = 0.5888\ldots$. Then $10x = 5.888\ldots$ and $100x = 58.888\ldots$. Subtract: $90x = 53$, so $x = \frac{53}{90}$.

3 $3^2 = 9$ and $4^2 = 16$. Since $9 < 15 < 16$, we have $3 < \sqrt{15} < 4$. Therefore $\sqrt{15} > 3$ is true. In fact, $\sqrt{15} \approx 3.87$.

4 $\sqrt{2} \times \sqrt{18} = \sqrt{2 \times 18} = \sqrt{36} = 6$.

5 $7 \times 7 = 49$, so $\sqrt{49} = 7$.

6 Move the decimal 3 places right: $6.1 \times 10^3 = 6{,}100$.

7 $\frac{2 \times 10^{-6}}{10^{-6}} = 2$. The bacterium is 2 micrometers long.

8 Taxi A: \$3/mile. Taxi B: $\frac{36}{9} = \$4$/mile. Taxi B is more expensive per mile.

9 $m = \frac{5-(-1)}{2-(-4)} = \frac{6}{6} = 1$.

10 $5x - 2x = 9 + 3$, so $3x = 12$, giving $x = 4$.

11 Rewrite the second: $-2y = -4x - 10$, so $y = 2x + 5$. Both equations are the same line. Infinitely many solutions.

12 $a + b = 12$ and $5a + 10b = 85$. From first: $a = 12 - b$. Substitute: $5(12 - b) + 10b = 85$, so $60 + 5b = 85$, $5b = 25$, $b = 5$.

13 Graph A is a circle and fails the vertical line test. Graph B is a parabola opening upward — every vertical line crosses it at most once, so it is a function.

Get Online

Find more at
ViewMath.com/RI-Grade8

ViewMath.com

14 $f(2) - f(1) = 7 - 4 = 3$.

15 *Function A earns \$12/week ($m = 12$). Function B earns \$15/week ($m = 15$). Since $15 > 12$, Function B earns more per week.*

16 *Graph Q is a straight line, which means the rate of change is constant. Graph P curves upward, meaning its rate of change increases.*

17 *No activation fee means $b = 0$. The monthly cost is \$35, so $m = 35$. The function is $y = 35x$.*

18 *Standing still means distance doesn't change: horizontal line. Then walking at a steady pace means distance increases at a constant rate: line sloping upward.*

19 *A rotation turns (spins) a figure around a fixed point called the center of rotation.*

20 *A dilation by a factor other than 1 changes the size of the figure, so it does not preserve congruence.*

21 *The $90°$ CCW rule is $(x, y) \to (-y, x)$. So $(-3, 5) \to (-5, -3)$.*

22 *When $k = 1$, every point stays in the same place. The image is congruent (identical in size and shape) to the original.*

23 *The $115°$ angle and x are co-interior (same-side interior) angles. They sum to $180°$: $x = 180 - 115 = 65°$.*

24 $c^2 = 10^2 + 24^2 = 100 + 576 = 676$, *so* $c = \sqrt{676} = 26$.

25 $d = \sqrt{(4-1)^2 + (7-3)^2} = \sqrt{9+16} = \sqrt{25} = 5$.

26 $V = \pi r^2 h$. *Doubling r: $V = \pi(2r)^2 h = 4\pi r^2 h$. The volume is 4 times as large because r is squared.*

27 Clusters are groups of points that bunch together, while outliers are isolated far from the trend. Both can appear in the same scatter plot.

28 Slope $= \frac{13-5}{6-2} = 2$. Using $(2,5)$: $5 = 2(2) + b$, $b = 1$. Line: $y = 2x + 1$.

29 Slope $= 0.4$ means 0.4 cm of growth per day. y-intercept $= 2$ means the initial height is 2 cm.

30 $0.15 \times 200 = 30$.

Practice Test 5 — Answer Key

1 Rational | 2 B | 3 A | 4 B | 5 C | 6 11 places to the left | 7 A | 8 D

9 B | 10 $x = 8$ | 11 $(3,7)$ | 12 B | 13 $\{(1,2),(2,3),(3,4)\}$ | 14 B | 15 A | 16 4

17 A | 18 C | 19 C | 20 C | 21 $(-2,5)$ | 22 C | 23 A | 24 B | 25 C

26 A | 27 Direction (positive/negative/none), shape (linear/nonlinear), and unusual features (outliers/clusters

28 C | 29 Predicted: $y = 1.5(6) + 2 = 11$. Residual $= 10 - 11 = -1$. | 30 B

Time to Learn!

Review the explanations below, **especially for the questions you missed**.

Understanding why each answer is correct builds stronger problem-solving skills.

Tip: Circle any questions you got wrong, then read their explanation carefully.

Practice Test 5 — Detailed Explanations

1 $\sqrt{144} = 12$ exactly, because $12^2 = 144$. Since 12 is an integer, $\sqrt{144}$ is rational.

2 The repeating block has 2 digits, so multiply by $10^2 = 100$. Then $100x = 63.6363\ldots$ and $x = 0.6363\ldots$. Subtracting: $100x - x = 63$.

3 $4^2 = 16$ and $5^2 = 25$. Since $16 < 22 < 25$, we have $4 < \sqrt{22} < 5$.

4 Area $= \sqrt{20} \times \sqrt{8} = \sqrt{160}$. $\sqrt{160} \approx 12.65$ cm^2. (Note: $\sqrt{160} = 4\sqrt{10} \approx 12.65$.) Choices A and C confuse addition with multiplication.

5 $81 = 9^2$, so 81 is a perfect square.

6 The exponent is -11, so you move the decimal 11 places to the left from 2.5

7 Same exponent, so add the coefficients: $5.2 + 3.8 = 9.0$. Answer: 9×10^6.

8 If the relationship passes through $(1, 5)$ and $(3, 15)$, then $k = 5$. But $5 \times 2 = 10 \neq 13$, so $(2, 13)$ does not fit.

9 $m = \frac{7-3}{\frac{3}{2}-\frac{1}{2}} = \frac{4}{1} = 4$.

10 $6x - 15 = 4x + 1$. Subtract $4x$: $2x - 15 = 1$. Add 15: $2x = 16$, so $x = 8$.

11 $2x + 1 = -x + 10$. So $3x = 9$, $x = 3$. Then $y = 2(3) + 1 = 7$.

12 $p + n = 8$ and $2p + 5n = 25$. From first: $p = 8 - n$. Substitute: $2(8 - n) + 5n = 25$, so $16 + 3n = 25$, $3n = 9$, $n = 3$.

Get Online
Find more at
ViewMath.com/RI-Grade8
ViewMath.com

13 *Any set where each input appears only once is correct. For example,* $\{(1,2),(2,3),(3,4)\}$ *has no repeated inputs.*

14 $f(0) = -1$ *and* $f(2) = 7$*. So* $f(0) + f(2) = -1 + 7 = 6$*.*

15 *Function A has slope* 6*. Function B:* $\frac{15-5}{3-1} = \frac{10}{2} = 5$*. Since* $6 > 5$*, Function A grows faster.*

16 $6 - 2 = 4$*,* $10 - 6 = 4$*,* $14 - 10 = 4$*. The rate of change is* 4*.*

17 *Slope* $= \frac{10-2}{3-(-1)} = \frac{8}{4} = 2$*.*

18 *When a graph falls from left to right, the output is getting smaller — the function is decreasing.*

19 *A* $180°$ *rotation uses the rule* $(x, y) \to (-x, -y)$*. So* $(-4, 3) \to (4, -3)$*.*

20 *Both rectangles have dimensions* 6×10*. A rotation maps one onto the other, so they are congruent.*

21 *Multiply each coordinate by* $\frac{1}{2}$*:* $(-4 \times \frac{1}{2},\ 10 \times \frac{1}{2}) = (-2, 5)$*.*

22 *Check ratios:* $\frac{6}{3} = 2$ *but* $\frac{9}{5} = 1.8$*. Since* $2 \neq 1.8$*, the sides are not proportional and the rectangles are not similar.*

23 *The exterior angle adjacent to the third angle equals the sum of the two non-adjacent interior angles:* $37 + 53 = 90°$*.*

24 $c^2 = 6^2 + 8^2 = 36 + 64 = 100$*, so* $c = 10$*.*

25 $d = \sqrt{(0-(-3))^2 + (0-(-4))^2} = \sqrt{9+16} = \sqrt{25} = 5$*.*

26 *Radius* $= 12$. $V = \frac{4}{3}(3.14)(12^3) = \frac{4}{3}(3.14)(1728) = \frac{4}{3}(5425.92) \approx 7234.6$ cm^3.

27 *Direction tells you how the variables relate. Shape tells if the trend is a line or curve. Outliers and clusters are notable features.*

28 *The better line balances points above and below while minimizing distances to data points.*

29 *The model predicts* 11*, but the actual is* 10*. The point is below the line by* 1.

30 *Of* 42 *pizza lovers,* 24 *are boys.* $\frac{24}{42} \approx 0.571$.

Practice Test 6 — Answer Key

1 Rational	2 C	3 B	4 B	5 A	6 B	7 A	8 B	9 A
10 C	11 B	12 C	13 No	14 C	15 C	16 D	17 $y = 4x + 4$	
18 Constant	19 C	20 30 cm	21 B	22 1,500 m	23 B	24 C	25 C	
26 B	27 B	28 B	29 B	30 B				

Time to Learn!

Review the explanations below, ***especially for the questions you missed.***

Understanding why each answer is correct builds stronger problem-solving skills.

Tip: *Circle any questions you got wrong, then read their explanation carefully.*

Practice Test 6 — Detailed Explanations

1 $\frac{5}{6}$ is a fraction of two integers with a nonzero denominator, so it is rational. Its decimal is $0.8\overline{3}$, which repeats.

2 The repeating block 123 has 3 digits, so you multiply by $10^3 = 1000$.

3 $10^2 = 100$ and $11^2 = 121$. Since $100 < 110 < 121$, we have $10 < \sqrt{110} < 11$.

4 $\sqrt{a} + \sqrt{b} \neq \sqrt{a+b}$. You must compute each root separately: $\sqrt{4} = 2$ and $\sqrt{9} = 3$, so the sum is 5. $\sqrt{13} \approx 3.6 \neq 5$.

5 Edge $= \sqrt[3]{343} = 7$ in., since $7^3 = 343$.

6 Move the decimal 5 places left: $9.03 \times 10^{-5} = 0.0000903$.

7 $4 \times (9.5 \times 10^{12}) = 38 \times 10^{12} = 3.8 \times 10^{13}$ km.

8 $k = \frac{y}{x} = \frac{9}{2} = 4.5$.

9 $m = \frac{1-7}{5-2} = \frac{-6}{3} = -2$.

10 Subtract 3: $\frac{x}{4} = 4$. Multiply by 4: $x = 16$.

11 $4x = -x + 15$. So $5x = 15$, $x = 3$. Then $y = 4(3) = 12$. Solution: $(3, 12)$.

12 $20x = 50 + 10x$. So $10x = 50$, $x = 5$. After 5 months the costs are equal; after more than 5 months Plan A costs more.

13 Input a maps to two different outputs (5 and 12), so the relation cannot be a function.

14 Set $5x + 2 = 27$. Subtract 2: $5x = 25$. Divide by 5: $x = 5$.

15 Function R: when $x = 0$, $y = 1$. Function S: $b = 1$. Both start at 1, so the initial values are equal.

16 $y = \sqrt{x}$ involves a square root of the variable, so it is nonlinear. The others all fit $y = mx + b$.

17 Slope: $\frac{24-4}{5-0} = \frac{20}{5} = 4$. y-intercept: 4. Equation: $y = 4x + 4$.

18 A straight line has a constant slope, meaning the rate of change is the same everywhere along the line.

19 Reflecting over the y-axis flips the sign of the x-coordinate while keeping y the same: $(7, -2) \to (-7, -2)$.

20 Congruent figures have equal perimeters because all corresponding sides are equal.

21 $180°$ rotation: $(a, b) \to (-a, -b)$. Reflect over x-axis: $(-a, -b) \to (-a, b)$.

22 Real length $= 3 \times 50{,}000 = 150{,}000$ cm $= 1{,}500$ m.

23 Corresponding angles are in the same position (e.g., both upper-left) at each intersection of the transversal with the parallel lines.

24 $c^2 = 9^2 + 12^2 = 81 + 144 = 225$, so $c = 15$.

25 $d = \sqrt{(4-(-4))^2 + (3-(-3))^2} = \sqrt{64+36} = \sqrt{100} = 10$.

26 $V = \pi r^2 h = \pi(4)(7) = 28\pi$ cm^3.

Get Online

Find more at
ViewMath.com/RI-Grade8

ViewMath.com

27 More exercise tends to lower resting heart rate. One goes up, the other goes down — negative association.

28 A horizontal line ignores the negative trend. The line of best fit should slope downward to match the data.

29 $y = -4(10) + 100 = -40 + 100 = 60$.

30 A two-way frequency table displays data for two categorical variables simultaneously, showing how frequently each combination occurs.

Practice Test 7 — Answer Key

1 $\sqrt{2}$ (or $\sqrt{3}$)
2 C
3 9 and 10
4 B
5 6 cm
6 B
7 317 times
8 C
9 B
10 B
11 $(2.5, 3)$
12 B
13 C
14 C
15 5
16 B
17 B
18 C
19 C
20 Reflection over the y-axis
21 C
22 B
23 $45°$
24 C
25 9
26 C
27 Negative linear association. As x increases, y decreases in a roughly straight pattern. No obvious outliers.
28 C
29 B
30 B

Time to Learn!

Review the explanations below, **especially for the questions you missed.**

Understanding why each answer is correct builds stronger problem-solving skills.

Tip: Circle any questions you got wrong, then read their explanation carefully.

Practice Test 7 — Detailed Explanations

1 $\sqrt{2} \approx 1.414$ *is between* 1 *and* 2 *and is irrational because* 2 *is not a perfect square.* $\sqrt{3} \approx 1.732$ *also works.*

2 $\frac{36}{100} = 0.36$ *(terminates). The correct conversion uses* $99x = 36$*, giving* $\frac{36}{99} = \frac{4}{11}$*. The student treated* $0.\overline{36}$ *as if it were* 0.36*.*

3 $9^2 = 81$ *and* $10^2 = 100$*. Since* $81 < 83 < 100$*, we have* $9 < \sqrt{83} < 10$*.*

4 $2\sqrt{5} = \sqrt{4} \cdot \sqrt{5} = \sqrt{4 \times 5} = \sqrt{20}$*. Alternatively,* $2\sqrt{5} \approx 4.47$ *while* $\sqrt{10} \approx 3.16$*. They are not equal.*

5 *Edge* $= \sqrt[3]{216} = 6$ *cm, since* $6 \times 6 \times 6 = 216$*.*

6 *Move the decimal* 5 *places left:* $350{,}000 = 3.5 \times 10^5$*. In scientific notation,* a *must satisfy* $1 \leq a < 10$*.*

7 $\frac{1.9 \times 10^{27}}{6 \times 10^{24}} = \frac{1.9}{6} \times 10^3 \approx 0.317 \times 10^3 = 317$*.*

8 *In choice C,* $\frac{9}{3} = 3$ *but* $\frac{15}{6} = 2.5$*. The ratio is not constant, so it is not proportional.*

9 $m = \frac{12-3}{4-1} = \frac{9}{3} = 3$*.*

10 *Subtract* 1.5*:* $0.5x = 2.5$*. Divide by* 0.5*:* $x = 5$*.*

11 *Subtract the second from the first:* $4y = 12$*, so* $y = 3$*. Then* $2x + 3 = 8$*,* $2x = 5$*,* $x = 2.5$*.*

12 $h + b = 80$ *and* $3h + 5b = 340$*. From first:* $b = 80 - h$*. Substitute:* $3h + 5(80 - h) = 340$*, so* $-2h + 400 = 340$*,* $-2h = -60$*,* $h = 30$*.*

13 *A function must give exactly one output per input. If button A yields two different results, the rule is violated.*

14 *From the table,* $p(0) = 1$ *and* $p(2) = 1$. *Both inputs give an output of* 1.

15 $\frac{19-4}{3-0} = \frac{15}{3} = 5.$

16 *A linear function has a constant rate of change (slope). Nonlinear functions have a rate of change that varies.*

17 *The rate is* \$3 *per mile* ($m = 3$) *and the flat fee is* \$4 ($b = 4$). *So* $y = 3x + 4$.

18 *A function can decrease while staying positive (e.g., dropping from* 10 *to* 5*). "Decreasing" means the values get smaller, not that they are negative.*

19 *A* $180°$ *rotation of a square around its center maps the square onto itself because of its symmetry.*

20 *Each* x*-coordinate is negated while* y *stays the same:* $(1, 1) \to (-1, 1)$, *etc. This is a reflection over the* y*-axis.*

21 *The rule* $(x, y) \to (-y, x)$ *is a* $90°$ *counterclockwise rotation around the origin.*

22 *Any circle can be mapped onto any other circle by a translation (to align centers) followed by a dilation (to match radii).*

23 $180 - 62 - 73 = 45°$.

24 *The ladder is the hypotenuse:* $c^2 = 12^2 + 9^2 = 144 + 81 = 225$, *so* $c = 15$ *ft.*

25 *Same* x*-coordinates, so distance* $= |7 - (-2)| = 9$.

Get Online

Find more at
ViewMath.com/RI-Grade8

ViewMath.com

26 $V = \frac{4}{3}\pi(3r)^3 = \frac{4}{3}\pi(27r^3) = 27 \times \frac{4}{3}\pi r^3$. *The volume is 27 times as large.*

27 *The points decrease steadily from* $(1, 10)$ *to* $(6, 2)$. *Direction: negative. Shape: approximately linear. No outliers or clusters.*

28 *A line of best fit (trend line) is a straight line drawn to approximate the overall trend, coming close to most points.*

29 *Predicted:* $y = 2.5(4) + 1 = 11$. *Residual = actual − predicted* $= 12 - 11 = 1$

30 *No association means the distribution of one variable does not change based on the other, relative frequencies should be roughly equal across rows.*

Practice Test 8 — Answer Key

1 C | 2 $\frac{81}{99} = \frac{9}{11}$ | 3 ≈ 6.7 | 4 D | 5 -5 | 6 C | 7 B | 8 A | 9 B

10 No solution | 11 C | 12 B | 13 Yes | 14 3 | 15 7 | 16 Linear | 17 \$25

18 B | 19 Translation, reflection, rotation | 20 C | 21 $(8, 0)$ | 22 A | 23 B | 24 Yes

25 10 | 26 C

27 *Yes. Example: population growth over time may curve upward (exponential), showing a positive but nonlinear association*

28 C

29 *At* $x = 5$: $y = 30$. *At* $x = 30$: $y = -45$. *The prediction at* $x = 5$ *is more reliable because it is within the data range;* $x = 3$

30 C

Time to Learn!

Review the explanations below, ***especially for the questions you missed.***

Understanding why each answer is correct builds stronger problem-solving skills.

Tip: *Circle any questions you got wrong, then read their explanation carefully.*

Practice Test 8 — Detailed Explanations

1 Every integer n can be written as $\frac{n}{1}$, making it rational. Not every square root is irrational (e.g., $\sqrt{4} = 2$).

2 Following the same pattern: $99x = 81$, so $x = \frac{81}{99}$. Simplify by dividing both by 9: $\frac{81}{99} = \frac{9}{11}$.

3 $6.7^2 = 44.89$ and $6.8^2 = 46.24$. Since 45 is very close to 44.89, $\sqrt{45} \approx 6.7$.

4 $\sqrt{3} \times \sqrt{12} = \sqrt{36} = 6$ and $\sqrt{6} \times \sqrt{6} = (\sqrt{6})^2 = 6$. But $\sqrt{3} + \sqrt{12} \approx 1.73 + 3.46 = 5.19 \neq 6$.

5 $(-5)^3 = -125$, so $\sqrt[3]{-125} = -5$.

6 $1 \times 10^2 = 100$ kg. From the chart, the lion's bar reaches close to the 10^2 mark, making its mass closest to 100 kg.

7 $\frac{6.4 \times 10^{10}}{4 \times 10^6} = 1.6 \times 10^4 = 16{,}000$ photos.

8 Machine A: 12 bottles/hour. Machine B: $\frac{50}{5} = 10$ bottles/hour. Machine A is faster.

9 $\frac{50-68}{6} = \frac{-18}{6} = -3$°F per hour. The negative sign shows the temperature is decreasing.

10 *Subtract $4x$: $7 = -3$, which is false. There is no solution.*

11 *The lines are parallel (same slope, different intercepts) and never cross. No solution.*

12 $c + w = 20$ *and* $2c + 4w = 56$. *From first:* $c = 20 - w$. *Substitute:* $2(20 - u) + 4w = 56$, *so* $40 + 2w = 56$, $2w = 16$, $w = 8$.

13 *For any value of x, there is exactly one value of x^2. Even though $(-3)^2$ and 3^2 both equal 9, each individual input still produces one output.*

14 *Follow the dashed line at $y = 6$ until it meets the graph. That point is $(3, 6)$, so $x = 3$.*

15 *The initial value is the output when $x = 0$, which is 7.*

16 *A constant increase of 6 means the rate of change is constant, which is the defining property of a linear function.*

17 $y = 5(3) + 10 = 15 + 10 = 25$.

18 *Accelerating: speed increases. Cruising: speed stays the same (constant). Braking: speed decreases.*

19 *The three rigid transformations are translation (slide), reflection (flip), and rotation (turn). They all preserve size and shape.*

20 *The triangles have the same angles, making them similar. But the base of Triangle A is 6 cm while Triangle B's base is 3 cm, so they are not congruent.*

21 $90°$ *CCW:* $(x, y) \to (-y, x)$. *So* $(0, -8) \to (8, 0)$.

22 *Check ratios:* $\frac{6}{4} = \frac{3}{2}$, $\frac{9}{6} = \frac{3}{2}$, $\frac{12}{8} = \frac{3}{2}$. *All equal, so the triangles are similar with* $k = \frac{3}{2}$.

Get Online
Find more at
ViewMath.com/RI-Grade8
ViewMath.com

23 *By the Exterior Angle Theorem:* $125 = 60 + x$, *so* $x = 65°$.

24 $11^2 + 60^2 = 121 + 3600 = 3721 = 61^2$. *Since* $a^2 + b^2 = c^2$, *it is a right triangle.*

25 $d = \sqrt{6^2 + 8^2} = \sqrt{36 + 64} = \sqrt{100} = 10$.

26 $V = \pi r^2 h = \pi(49)(3) = 147\pi\ m^3$.

27 *An association can be positive (both increase) and nonlinear (the trend is curved, not straight).*

28 *The* y*-intercept is the* y*-value when* $x = 0$*. The line passes through* $(0, 3)$*, so* $b = 3$.

29 $y = -3(5) + 45 = 30$ *(interpolation).* $y = -3(30) + 45 = -45$ *(extrapolation — unreliable and gives a negative value that may not make sense).*

30 $\frac{25}{40} = 0.625 = 62.5\%$.

Practice Test 9 — Answer Key

1 C
2 B
3 A
4 C
5 B
6 B
7 C
8 10 packages
9 A
10 $x = \frac{5}{3}$
11 B
12 B
13 B
14 C
15 B
16 C
17 A
18 A
19 C
20 B
21 B
22 C
23 B
24 B
25 B
26 B
27 C
28 C
29 B
30 *Male Left* $= 5$; *Female Right* $= 32$; *Total Left* $= 13$; *Total Right* $= 67$.

Time to Learn!

*Review the explanations below, **especially for the questions you missed.***

Understanding why each answer is correct builds stronger problem-solving skills.

***Tip:** Circle any questions you got wrong, then read their explanation carefully.*

Practice Test 9 — Detailed Explanations

1 $1.41421356\ldots$ is the decimal expansion of $\sqrt{2}$, which is non-repeating and non-terminating, indicating an irrational number.

2 Let $x = 0.444\ldots$. Then $10x = 4.444\ldots$. Subtract: $9x = 4$, so $x = \frac{4}{9}$. Note that $\frac{4}{10} = 0.4$ (terminates), which is different from $0.\overline{4}$.

3 $6^2 = 36$ and $7^2 = 49$. Since $36 < 45 < 49$, we have $6 < \sqrt{45} < 7$. The student is correct.

4 $\sqrt{2} \approx 1.414$ and $\sqrt{3} \approx 1.732$, so their sum is approximately 3.146. $\sqrt{10} \approx 3.162$. Since $3.162 > 3.146$, $\sqrt{10}$ is slightly larger.

5 $7^2 = 49$ and $8^2 = 64$. Since $49 < 50 < 64$, we have $7 < \sqrt{50} < 8$.

6 $8 \times 10^5 = 800{,}000$ and $3 \times 10^6 = 3{,}000{,}000$. The higher exponent wins: 3×10^6 is larger.

7 $7 \times 8 = 56$ and $10^{-3} \times 10^{-2} = 10^{-5}$. Then $56 \times 10^{-5} = 5.6 \times 10^{-4}$.

8 Service P: $4 \times 10 = 40$ packages. Service Q: $k = \frac{10}{2} = 5$, so $5 \times 10 = 50$ packages. Difference: $50 - 40 = 10$.

9 $m = \frac{7-(-2)}{3-0} = \frac{9}{3} = 3$.

10 *Left side:* $3x + 4$*. Right side:* 9*. So* $3x + 4 = 9$*, giving* $3x = 5$ *and* $x = \frac{5}{3}$.

11 *Subtract the second from the first:* $2x = 8$*, so* $x = 4$*. Then* $4 + 2y = 8$*, so* $2y = 4$*,* $y = 2$.

12 $a + c = 5$ *and* $10a + 6c = 38$*. From first:* $c = 5 - a$*. Substitute:* $10a + 6(5 - a) = 38$*, so* $4a + 30 = 38$*,* $4a = 8$*,* $a = 2$.

13 *Input* 3 *appears twice with outputs* 9 *and* 15*. Since one input maps to two different outputs, the relation is not a function.*

14 $g(3) = 3^2 - 5 = 9 - 5 = 4$.

15 *Plan X:* $5x + 40$*. Plan Y:* $10x + 10$*. Set equal:* $5x + 40 = 10x + 10 \Rightarrow 30 = 5x \Rightarrow x = 6$.

16 $y = -2x + 6$ *fits the form* $y = mx + b$ *with* $m = -2$ *and* $b = 6$*. A negative slope does not make a function nonlinear.*

17 *Slope:* $\frac{22-10}{5-2} = \frac{12}{3} = 4$*. Using* $(2, 10)$*:* $10 = 4(2) + b \Rightarrow b = 2$*. So* $y = 4x + 2$.

18 *Walking increases distance. Waiting means distance stays the same (constant). Walking again increases distance further.*

19 *Reflections, rotations, and translations are rigid transformations — they preserve size and shape. Dilations and stretches change size.*

20 *Two rigid transformations are described: one reflection and one translation.*

21 *Original sides:* 6*,* 8*, and* $\sqrt{6^2 + 8^2} = 10$*. Perimeter* $= 6 + 8 + 10 = 24$*. Dilation by* $\frac{1}{2}$ *multiplies every length by* $\frac{1}{2}$*, so the new perimeter is* $24 \times \frac{1}{2} = 12$.

22 *Similar figures can be mapped by a combination of rigid transformations (translations, reflections, rotations) and dilations.*

23 *Alternate interior angles are equal:* $3x + 10 = 5x - 30$*, so* $40 = 2x$ *and* $x = 20$.

24 *The Pythagorean Theorem states that in a right triangle,* $a^2 + b^2 = c^2$ *where* c *is the hypotenuse.*

25 *Both points have* $y = 2$*. Distance* $= |7 - 1| = 6$.

26 $200\pi = \pi r^2(8)$*, so* $r^2 = \frac{200}{8} = 25$*, and* $r = 5$ *cm.*

27 *Point* P *at* $(3, 7)$ *is far above the general upward trend around* $y \approx 2.5$ *when* $x = 3$*. It is the outlier.*

28 *In* $y = mx + b$*, the slope is* $m = -3$.

29 *Residual = actual − predicted. A positive residual means the actual value is above the predicted value.*

30 *Male Left* $= 40 - 35 = 5$*. Female Right* $= 40 - 8 = 32$*. Total Left* $= 5 + 8 = 13$*. Total Right* $= 35 + 32 = 67$.

Practice Test 10 — Answer Key

1 B | 2 D | 3 B | 4 D | 5 C | 6 Cell C, Cell A, Cell B | 7 3×10^6 | 8 99

9 B | 10 D | 11 $(2, 5)$ | 12 C | 13 $0, 3, 6$ | 14 A | 15 B | 16 Linear | 17 A

18 0–4 min: linear increasing; 4–8 min: constant; 8–12 min: nonlinear decreasing. | 19 B | 20 C

21 B | 22 C | 23 83° | 24 C | 25 C | 26 ≈ 226.08 in^3 | 27 B | 28 A | 29 A

30 B

Time to Learn!

*Review the explanations below, **especially for the questions you missed.***

Understanding why each answer is correct builds stronger problem-solving skills.

***Tip:** Circle any questions you got wrong, then read their explanation carefully.*

Practice Test 10 — Detailed Explanations

1 *5 is not a perfect square, so $\sqrt{5}$ is irrational and belongs in the Irrational region.*

2 *In Step 5, the student divided 2 by 10 instead of 9. From $9x = 2$, the correct answer is $x = \frac{2}{9}$.*

3 $7^2 = 49$ *and* $8^2 = 64$. *Since* $49 < 50 < 64$, *we have* $7 < \sqrt{50} < 8$.

4 $\pi \approx 3.1416$, $\sqrt{10} \approx 3.1623$, *and 3.2 is exact. From least to greatest:* $3.1416 < 3.1623 < 3.2$.

5 $\sqrt{2}$ *cannot be expressed as a fraction of two integers, so it is irrational.*

6 *Cell A:* 8×10^{-3} *mm. Cell B:* 5×10^{-2} *mm. Cell C:* 3×10^{-4} *mm. From smallest to largest:* $3 \times 10^{-4} < 8 \times 10^{-3} < 5 \times 10^{-2}$, *so Cell C, Cell A, Cell B.*

7 $\frac{3.6}{1.2} = 3$ *and* $10^{10-4} = 10^6$. *Answer:* 3×10^6.

8 $k = \frac{54}{6} = 9$. *When* $x = 11$: $y = 9 \times 11 = 99$.

9 *A line that goes down from left to right has a negative slope.*

10 *Subtracting $6x + 4$ from both sides gives $0 = 0$, which is always true. Every value cf x is a solution.*

11 *Both tables give $y = 5$ when $x = 2$. The solution is $(2, 5)$.*

12 $2l + 2w = 34$ *and* $l = w + 5$. *Substitute:* $2(w + 5) + 2w = 34$, *so* $4w + 10 = 34$, $4w = 24$, $w = 6$.

13 *The inputs are the values in the left column of the mapping diagram: 0, 3, and 6.*

14 $f(10) = 7 - 10 = -3$.

15 $G(4) = 14$. $H(4) = 4(4) + 1 = 17$. *Since* $17 > 14$, *Function H has the greater value at* $x = 4$.

16 *It fits* $y = mx + b$ *with* $m = 8$ *and* $b = 1$. *The variable has exponent 1, so it's linear.*

17 *Slope:* $\frac{16-7}{6-3} = \frac{9}{3} = 3$. *Using* $(3, 7)$: $7 = 3(3) + b \Rightarrow 7 = 9 + b \Rightarrow b = -2$.

18 *The first section is a straight line going up (linear increasing). The middle section is flat (constant). The last section curves downward (nonlinear decreasing).*

19 *Translations preserve all side lengths. The longest side is still 8 cm.*

20 *Congruent figures have identical side lengths and angle measures. They do nct need to be in the same position or orientation.*

21 *Add the translation to each coordinate:* $(5 + (-2),\ -3 + 4) = (3, 1)$.

Get Online
Find more at
ViewMath.com/RI-Grade8
ViewMath.com

22 *All squares have four 90° angles. Any two squares can be mapped by a dilation since the ratio of their sides is constant. So all squares are similar.*

23 *Co-interior angles are supplementary:* $180 - 97 = 83^\circ$.

24 $6^2 + 8^2 = 36 + 64 = 100$, *but* $11^2 = 121$. *Since* $100 \neq 121$, *this is not a right triangle.*

25 $d = \sqrt{8^2 + 6^2} = \sqrt{64 + 36} = \sqrt{100} = 10$.

26 $V = 3.14 \times 9 \times 8 = 226.08\ in^3$.

27 *When one variable increases and the other decreases, it is a negative association.*

28 *The data has a positive trend. Line A follows this trend with roughly equal points above and below. Line B is flat and ignores the upward pattern.*

29 $y = -1{,}500(5) + 20{,}000 = -7{,}500 + 20{,}000 = 12{,}500$.

30 85.7% *of students who studied passed vs.* 40% *who did not study. The large difference suggests an association.*

Well done checking your answers!

Keep practicing to strengthen your skills.